Pancreatic Cancer:
It's a Family Affair

Pancreatic Cancer: It's a Family Affair

Lisa M. Strahs-Lorenc

To order additional copies of this book, contact:
Xlibris Corporation
1-888-795-4274
www.Xlibris.com
Orders@Xlibris.com
83838

CONTENTS

DEDICATION

To my very brave husband with whom I shared thirty-three years of love and to our children, I hope you will always remember the importance of family and pass on this value to generations to come.

This book is not meant to be a medical guide for Pancreatic Cancer. We are neither doctors nor healthcare providers. The comments and advice in this book are based solely on our personal experiences, and should not be taken as medical advice or recommendations.

FOREWORD

This is not another book about cancer. One can go to a bookstore or library and find many books detailing the disease and the myriad of ways to "attack" it—traditional medicine, nontraditional medicine, and no medicine at all. This is not a book about the disease. Rather, it is a book about the effects of the disease on the patient, the spouse or partner, the children, the parents, and other relatives. It is a book about the effects of cancer on the family. Cancer affects the whole family, forever changed and forever coping with the disease. While there are support groups for patients, caregivers, and children, the family unit—its dynamics, its stability, its foundation—will never be the same. The families who have contributed their stories, their feelings, and their beliefs have done so in order to help others feel that they are not alone. This book is divided into sections, from "Before Cancer" (B.C.) to "Building a New Life," so that family members can share their experiences that will provide support to readers wherever they are on this sad journey.

The first chapter gives us a picture of life before pancreatic cancer—for some, a busy life, flying by; for others, a day-to-day, live-for-the-moment lifestyle; and for others, an uneventful lifestyle.

The second chapter is the big moment: the diagnosis. Some families experience the shock and disbelief so commonly described in books, and others suspect the seriousness of the disease that had gone unchecked for months.

In the third chapter, families describe various treatments and life after diagnosis. Each family copes differently with the effects of pancreatic cancer and the changes created for the patients and to their overall lives.

The fourth chapter describes how families deal with the reality that treatments are no longer working. The battles may have been won, but the war is being lost. Though treatments have been ongoing, the larger picture is that the disease has continued to spread. The patient and his/her family members feel that they are on a roller-coaster ride, not knowing what is going to happen next and always feeling that there is a cloud over their heads.

The fifth chapter deals with death and grieving. Families share their reactions, ways of coping, and ceremonies to celebrate the life of the loved one they lost.

In the last chapter, families share their lives as they move on without their loved one. Though forever changed, families find ways to build a new life.

Finally, the "Addendum: Advice from the Battlefield" offers advice to the reader and a look back at what they experienced. It also contains survivor updates.

In addition to providing support for family members, the author plans to donate all proceeds from any profit to the Lustgarten Foundation and Pancreatic Cancer Action Network, two organizations that continue to do research, advocacy, and offer support for pancreatic cancer. The book will also be distributed at no charge to the American Cancer Society, Cancer*Care*, and any other support group that offers support to any family member who has been affected by this dreadful disease.

ABOUT THE LUSTGARTEN FOUNDATION

The Lustgarten Foundation, based in Bethpage, New York, is America's largest private foundation dedicated solely to funding pancreatic cancer research. Founded in 1998, the foundation provides critical support in the search for better diagnostics and treatment of pancreatic cancer and, to date, has provided more than $32 million to more than 115 research projects at forty-one medical and research centers worldwide.

The Lustgarten Foundation and Cablevision Systems Corporation, a leading media and entertainment company, together launched curePC, a public awareness campaign that uses Cablevision's high-profile assets to draw attention to the fight against pancreatic cancer. As part of this campaign, Cablevision made a multiyear commitment to underwrite all the Lustgarten Foundation's administrative costs to ensure that 100 percent of every dollar donated to the foundation will go directly to pancreatic cancer research. To learn more, visit www.curepc.org.

ABOUT PANCREATIC CANCER ACTION NETWORK

Founded in 1999 by three visionaries as a small, nonprofit, 501(c) (3) patient-based advocacy organization, the Pancreatic Cancer Action Network understands the many challenges that patients and their families face in the fight against pancreatic cancer.

With our national headquarters in El Segundo, California, and a Government Affairs office in Washington DC, the organization fulfills its mission through a nationwide network of people dedicated to working together to advance research, support patients, and create hope for those affected by pancreatic cancer.

Under the direction of President and CEO Julie Fleshman, JD, MBA, the Pancreatic Cancer Action Network comprises of a national Board of Directors, a Scientific Advisory Board, a Medical Advisory Council, and a staff dedicated to fighting the disease.

The organization takes a unique and comprehensive approach to fighting pancreatic cancer. The Pancreatic Cancer Action Network is fully committed to providing the most comprehensive set of programs and services to fight this very difficult disease. Our commitment and passion are driven by our mission: to advance research, support patients, and create hope for anyone facing pancreatic cancer. To learn more, visit www.pancan.org.

ACKNOWLEDGMENTS

Kerri Kaplan, Executive Director, The Lustgarten Foundation

The Lustgarten Foundation thanks Lisa M. Strahs-Lorenc, BS, MPS and all the contributors to *Pancreatic Cancer: It's a Family Affair* for donating the proceeds of this book to the Lustgarten Foundation. The Lustgarten Foundation, based in Bethpage, New York, is America's largest private foundation working solely on pancreatic cancer research. The foundation provides critical support in the search for better diagnostics and treatment of pancreatic cancer. To learn more, visit www.lustgarten.org.

Julie Fleshman, President and CEO, Pancreatic Cancer Action Network

The Pancreatic Cancer Action Network applauds Lisa M. Strahs-Lorenc, BS, MPS for sharing her personal perspective on the affects of pancreatic cancer in her new book, *Pancreatic Cancer: It's A Family Affair*. I lost my own father to pancreatic cancer just four months after his diagnosis. My family and I know all too well how this disease devastates the patient and their entire family. The Pancreatic Cancer Action Network appreciates Ms. Strahs-Lorenc's generosity in donating proceeds of this book to the organization. The Pancreatic Cancer Action Network is the only national organization working together to advance research, support patients and create hope for all those affected by pancreatic cancer. To learn more and get involved, visit www.pancan.org

SPECIAL THANK-YOUS

I would like to extend a special thank-you to Rosanne K. Silberman, EdD, professor and coordinator, Programs in Blind and Visually Impaired and Severe/Multiple Disabilities including Deafblindness, Hunter College, CUNY. Her time—and detail-oriented editing of this book was invaluable.

As with all of us, her husband, Joe, had pancreatic cancer, which he battled for four years and then lost in January 2008. Since life does go on, Rosanne gets great pleasure from her kids and grandkids. She actually has five grandkids—Rebecca, five and a half; Jackie, four; and one more that was born during the writing of this book. They belong to her son, Marc, and daughter-in-law, Amy. She also has Max who is almost three and Eli Joseph (named after Joe) who is seven months old. They belong to Michelle (her daughter) and Clark.

In addition, I want to recognize the artistic talents of Sandra Y. Peyton, daughter-in-law to Jeanine (one of the contributors and spouse of one of the deceased patients). Her dedication and inspiring book cover and back have captured the essence of the message of this very personal book. I am so grateful for her important contribution to this book.

Where are the contributors from?

State	Number of contributors
California	3
Florida/Massachusetts (summer)	3
Illinois	1
Louisiana	1
Minnesota	2
New Jersey	1
New York	6
North Carolina	1
Oregon	1
Pennsylvania	1
Utah	1
Virginia	1

Ages of the patients

Age	Amount of patients
50–55	5
56–60	3
61–65	5
66–70	5
71–75	3

Contributors' relationship with patient

Caregiver-spouse	9
Caregiver-sibling	2
Son/daughter	9
Patient	2

FAMILIES*

FAMILIES	FAMILY #1	FAMILY #2	FAMILY #3 Note: Family #3 has 2 patients—father and mother. They had been divorced since 1976.	FAMILY #4	FAMILY #5	FAMILY #6
Patient first name	Fred	Charles	Ethel and Brian	Joycelyn, "Joyce"	Jenifer	Joe
Diagnosis and age at diagnosis	Adenocarcinoma, inoperable, age 55	Adenocarcinoma, age 68	Ethel diagnosed with pancreatic cancer during her Whipple surgery, age 70. Brian was diagnosed in December 2002 with pancreatic cancer at age 65.	Pancreatic cancer, age 51	Adenocarcinoma, tail of pancreas, age 53	Adenocarcinoma, age 59
Date of diagnosis	October 13, 2007	February 3, 2009	Ethel was diagnosed on January 4, 2007.Brian was diagnosed on December 2001.	July 13, 2007	September 11, 2006	November 1999
Status (date of death or treatments)	died January 30, 2009, had chemo and radiation	Attempted Whipple, unable to remove tumor, currently on Gemcitabine	Ethel died December 1, 2008. Brian died November 17, 2003.	Died October 9, 2009	No evidence of cancer since November 2008-on oral Tarceva daily	Whipple, chemo and radiation following
Contributor and relation to patient	**Lisa, wife and Ryan, son**	**Pat, wife**	**Jennifer R., daughter**	**Jackie, sister**	**Jenifer, self**	**Joe, self**
Significant family members			Jim (primary caregiver), married 27 years to Ethel	Gary, husband	Parents deceased	Myra, wife, age 66
Children	Sari, daughter, now 24; Ryan, son, now 20	Teresa	Jennifer R., daughter,(Brison, husband); Brian, Jr., son (Melissa, wife)	David, son, age 24; Vickie, daughter, 18	Katie, 24; Michael, 22; Brooke, 20; Megan, 17	Bruce, age 42, married to Nan; Wendy, age 40, married to Kevin
Siblings	Pat and Sharon		Sonya, sister, and Fred, brother-in-law, both deceased	Janet and Jackie, sisters; Ted, brother	two brothers, ages 59 and 53	Ed, age 61, married
Grandchildren/ other	None	HeatherLuay, son-in-law	Eden, age 2; Hannah, age 7; Jake, age 3; Luke, age 3; Robyn, nephew; Randi, grandniece; Travis, grandnephew	None	Great friend Deborah Barfield, POA, Healthcare Surrogate	Mother, still alive;Alex, Evan, Ethan, grandchildren

FAMILIES	FAMILY #7	FAMILY #8	FAMILY #9	FAMILY #10	FAMILY #11	FAMILY #12
Patient first name	Vincent	Jim	Greg	Dolly	Al	Don
Diagnosis and age at diagnosis	Pancreatic cancer, stage 3, diagnosed at age 71	October 18, 2005, diagnosed at age 51	Adenocarcinoma, diagnosed at age 58	Pancreatic cancer, stage 4 with metastasis to liver, age 63	Adenocarcinoma, stage 4, age 62	Adenocarcinoma , age 61
Date of diagnosis	August 2008	October 18, 2005	June 2008	January 2009	December 2006	June 29, 2006
Status (date of death or treatments)	tumor is stable, radiation completed, June 2009	died April 18, 2009	died February 27, 2009	Died on August 10, 2009	died June 23, 2008	died June 23, 2008
Contributor and relation to patient:	**Kathy, wife**	**Judy, wife**	**Michele, daughter and best friend**	**Deborah, daughter of Dolly**	**Beverly, wife**	**Jeanine, wife**
Significant family members						
Children	Lorraine, William, Michael, Jeannine, Robert, Bianca, Barbara Ann	no children	Michele	Nick, son	One son	Cheri, daughter; Lee, son
Siblings	Ann	One brother—John 57; two sisters—Joyce, 49; Janet, 52	Chris, brother; Ruth Marie and Nina, sisters	Irene, Susanna, Wendy, David	Deceased in 1989 from pancreatic cancer	Ron, David, and Jeff, brothers (all still living)
Grandchildren/ other	Nicole, Joseph, John, Michael, Michael,Jacqueline, William, Antonia, Robert, Alexandra, Robert	Frances, mother, age 78	Parents deceased	Kwan-Cha, father, Shu-Sha, mother	Mother deceased from breast cancer, father from prostate cancer	Ken and Virginia (both deceased before Don), Devin and Emma; Jessica

FAMILIES	FAMILY #13	FAMILY #14	FAMILY #15	FAMILY #16	FAMILY #17	FAMILY #18
Patient first name	Floyd	James	Margie	Len	Gary	Frayda
Diagnosis and age at diagnosis	Adenocarcinoma with metastasis to the liver, age 66	Pancreatic cancer, age 63 at diagnosis	Adenocarcinoma stage IV with liver metastasis; age 71	Pancreatic cancer, age 59	Adenocarcinoma stage III, age 54	Adenocarcinoma of the tail of the pancreas, with metastasis in the liver, age 70
Date of diagnosis	July 7, 2006	April 21, 2006	October 2006	Dec 18, 2005	April 1996	January 2008
Status (date of death or treatments)	died January 11, 2007	died April 16, 2008	died February 13, 2009	died May 12, 2006	died June 1997	died April 2, 2009
Contributor and relation to patient	**Bette, wife, and Carrie, daughter**	**Melissa, daughter**	**Sherri, daughter**	**Sue, wife**	**Mary, sister**	**Aaron, husband**
Significant family members					Survived by his wife	

Children	Floyd III, Christopher, Craig and Carrie	Melissa, daughter; James, son and Kirk, son-in-law; Cindi, daughter-in-law	Jimmy, Gary, and Sherri (contributor)	son David (age 29 at the time), his wife, Mel; daughter, Missy, age 24 at the time	Two daughters	Jeffrey, 51; Mark, 49; Laurie, 47
Siblings	Elizabeth, sister		Donnie and Katherine, sisters and Earl, brother, (died 4/12/09 from cancer after Mom died)	Wylma, Ora, Karen, Mikel	Four sisters (one predeceased)	Vivian, 73
Grandchildren/other	Floyd and Mildred, parents both deceased; seven grandchildren—Brandon, Shane, Jeremy, Tristan, Tiana, Beau, and Camdyn	Thelma and Warren, parents both deceased;Hayley and Aiden, grandchildren	Perry and Blanch, parents both deceased. Tony, grandchild	Parents both deceased, one granddaughter, Nichelle, age 5 at the time	Both parents survived him. Three grand-daughters, three grandsons, one stepgrandson	Parents Lillian—passed away at age 99—and Frederick, passed away at age 67; seven grandchildren, ages 7 to 20;Aunt Florence, who was very close to Frayda. She passed away about 15 years ago

FAMILIES	FAMILY #19	FAMILY #20
Patient first name	Stanley	Michele
Diagnosis and age at diagnosis	Adenocarcinoma, age 70	Adenocarcinoma, 73 years old
Date of diagnosis	January 2008	July 20, 2007
Status (date of death or treatments)	Treatment began April 2008, died September 2008	died January 25 2008
Contributor and relation to patient	Cinzia, step-daughter	Ernesto, son
Significant family members		Anna, spouse
Children		Maria Concetta, Armandina, Ernesto
Siblings		Nicola, brother; Gioconda, sister;Armando, brother, December 1942, killed in action;Fausta, sister, December 1989, breast cancer
Grandchildren/other		Parents Ernesto, Maria Concetta, both deceased; father from leukemia;Alyssa, Michelle, Michael, grandchildren

Note: Each family is described with the patient's first name, present age, patient diagnosis, and status, in addition to contributor and significant family members.

Highlighted patients are still alive as of the completion of this book.

*Bolded descriptions represent contributors to the book.

CHAPTER 1

B.C.: Before Cancer

Yesterday is history. Tomorrow is a mystery. And today? Today is a gift. That's why we call it the present.

—Babatunde Olatunji

This chapter is devoted to our lives before pancreatic cancer entered into our families.

As we think back about our lives, we discuss the relationship we had with family members, what brought us happiness and what we enjoyed, how we spent our free time, and what was our philosophy of life.

The relationship with family members is described below with a few common themes:

Very close families

Jackie describes her relationship with her family as "always very close, including spending time together on all major holidays and at other times as often as possible."

Aaron adds, "All of our family members, our three children, seven grandchildren, and three spouses of our children, all got along very well with each other."

Pat speaks about close family relationships that even distance doesn't change. "We had just returned from six weeks abroad, having spent ten days in Italy, followed by thirty-three days on a cruise ship. Before we left on that trip, our daughter, who lives on the opposite coast from us, had brain surgery, and we spent a month there helping take care of her and her family. The previous paragraphs probably define a large part of our lives. Our daughter and her family are such a big part of our lives even though we no longer live close to each other. When we moved away, we would tell people, 'That's why they make airplanes.' We spent a month or more at her home every Christmas and managed to get together with her and her family somewhere at least one other time every year. Our granddaughter is absolutely the light of our life. There is very little extended family, but we have reconnected with cousins recently—they are scattered all over the country, but we talk on the phone and exchange e-mails."

Cinzia explains her feelings about family. "Family is very important in my life. Finding time for dinners and celebrations is vital to living my Italian heritage and keeping the close bond with family members. Studying Italian is also part of my daily activities."

Bette elaborates on her relationship with her husband. "Floyd and I met in the seventh grade and dated ever since. We married in September 1960. Our family did most everything together and remained close even after acquiring in-laws and grandchildren. Floyd and I spent the last ten of his years doing family child care together in our home, which is what I previously did."

Jenifer M. writes that "we always had a very close family bond. I was on good terms with my ex-husband and his wife. My mother was alive and doing well for an eighty year old. We often had many family get-togethers."

Joe states that "our relationship with the kids has always been good. Prior to the cancer, we spent a lot of time with our first grandchild and were waiting for the second one to be born. We were also planning a wedding for our daughter."

Kathleen states that "Vinny has seven children—five from one marriage and two from a second. At the time we met, his youngest child was about eight years old, and she and her sister lived in Virginia and still do. Because of the distance, we only saw them four to five times a year, and even though our relationship with them was fine and without conflict, it was not particularly close. The other five children range in age from about thirty-five to twenty-two, and they live twenty miles away, thus nearby. As our dating relationship grew, I came to know and like all five children, and at that time, there were about eight grandchildren. I had tried to have children in my previous marriage, which had ended in divorce eight years previously, and was always saddened that I could not. To become part of this family was both wonderful and, at times, overwhelming. Suffice to say, I learned a great deal and felt that by the time Vinny and I married two and a half years later, I was accepted and loved."

Judy writes that "Jim's relationship with his siblings, mom, nieces, nephews, extended family was very good. Jim was well loved by *many* people and we both enjoyed just hanging around with siblings, friends, extended family, and others."

Michele describes her relationship with her dad as "the best you can imagine. We were best friends, and he was an inspiration to me—my mentor, my cheerleader, my foundation. He was the best dad in the world. I would call him for advice, just to check in, when my dog would do something funny, when I had a bad day at work. I talked to him every day, sometimes multiple times a day. Our father-daughter relationship was stronger than anyone we knew. We were a team and always had been. My mother passed away when I was seven, so my father raised me on his own. He joked around about how he just used to 'feed me and drive me places,' but the reality was that he was an incredible father with a deep wisdom on how to raise a daughter. I lived in San Diego, where I went to college and started a career, and he was in San Jose where I grew up. He asked me to move home every day."

Ryan has similar thoughts. "I learned so much from my dad, from his phenomenal work ethic to how well he treated the people he knew. My childhood was the best childhood that a kid could ever ask for because of the undying love that was shown to me. Ever since I can remember, my father was one of the most unique men that I had ever met. What made him such a unique person is the way that he handled the challenges in his life, while still maintaining a great relationship with my mom, sister, and me."

Deborah recounts the same kind of relationship with her mom. "I have been blessed to have an amazing relationship with my mother. She and I have only become closer and closer since I returned to the East Coast after attending college in California. I live in NYC, and my parents live in NJ, and I always treasured the Sundays when my mother would drive up to visit me. She would bring a car loaded with groceries, home-cooked food, and toilet paper (all of which are significantly less expensive in suburbia), and I would treat her to a massage at my local spa and dinner at a restaurant of her choice. My father would join her on occasion, but he has severe back pain and is often limited by the drive from NJ to NYC. In fact, my mother was actually *his* caregiver for many years now. They are both retired, and she would often complain about lacking a traveling companion since he is not able to handle long airplane rides. My mother actually took a long trip to California and Taiwan just last October, where she hiked up mountainous regions!

"I would return home on weekends to see my parents perhaps once every four to eight weeks, and my mother would make the drive up about as often. I would call home about once a week or whenever I had big news. Before my mother's diagnosis, my life was extremely self-oriented: I would see them whenever was convenient for me, I would go home and expect home-cooked food and a clean room, and I pretty much had nothing to worry about but myself. (I am unmarried, with no children and no house). One of my mother's great joys was spending time with my brother's children. She would take every occasion to drive the three hours out to his house to see them."

Sherri describes her relationship with her mom: "My mom and I were very close. She separated from her second husband in September of 2000. I was already divorced and living alone (no children but two dogs, a boxer named Zach and a Pomeranian named Raider who were literally my babies). Mom needed somewhere to live, so she moved in with me. She had said in the beginning it was only supposed to be temporary, but we never had any problems living together. Sure, I have friends, but Mom became the person I did things with and told things to, not just because I had to but because I wanted to. If you've ever heard the expression 'Nobody loves you like your momma,' that's how I felt (and still do). We always got together as a whole family (aunts, uncles, cousins, etc.) at major holidays and often saw each other, some of us anyway, on a weekly or at least monthly basis."

Another close relationship with parents is described through Carrie's words. "Life before cancer was a wonderful gift. It brought joy, peace, love, adventure, and security to my life. I felt so fortunate to have my parents and siblings close to me in vicinity and in daily life. At the age of thirty-seven, I knew it wouldn't last forever and cherished all the memories made together, never taking these times and days for granted. Growing up the youngest of four children and being the only daughter, I was a true daddy's little girl. Both my parents were very supportive and provided opportunity for my brothers and myself to experience a variety of activities, along with the security of unconditional love. Growing up, my dad was even awarded the Biggest Fan award at my high school track banquet because he never missed a meet. I might not have been the fastest runner, but I felt so proud when I would look over at my dad at the end of a race, and he would give me the thumbs-up for a job well done. It was the same after every piano competition, speech, and finally at college graduation. The love I felt growing up made me feel as if I won the lottery with my parents. Sometimes, I would be at a family gathering, and I would close my eyes to take a mental picture in my mind of the secure and special moments these times brought to forever store in my mind . . . knowing they couldn't last forever."

Ernesto explains his relationship with his father. "My vivid memories as a child are of him fixing VCRs, air conditioners, anything that needed fixing, even the family cars. I remember how amazed I felt when he rebuilt the family car engine. I thought, and still think, how amazing it is, considering he was a plumber by trade, a third-generation plumber without any formal schooling. Most of the time I spent alone with my dad growing up was going to work with him. After school and on Saturdays, I would go do side jobs with him. As a teenager, it became my job. I learned a lot and will forever be grateful for all that he taught me. Once I became older and went to college and later married, I did not have the time to go to work with him. That changed in 2007. By this time, my father had retired, and I was in the middle of an overwhelming home extension."

Jeanine describes her close relationship with her husband. "Don and I were devoted to each other to the point that my mother once warned me that I loved Don too much. We often talked about how much we enjoyed our closeness and dependency upon each other but knew that it would be hard on the one left if one of us died. However, we decided that whoever was left would have to deal with it if (we really hoped we'd go together) it happened and that we'd enjoy our relationship to the maximum while we had it. Of course, we felt certain we wouldn't have to worry about death for at least ten more years, so we were comfortable with that conclusion.

"Both our children were unplanned, coming before we felt ready to have them. Especially Don, as he wanted to be more financially stable before starting a family. However, we loved them deeply and were thankful when they became adults without any major mishaps. We had great friendships with them as adults and spent as much time with them as was possible. Both of us doted on our two grandchildren. One of Don's biggest regrets about leaving this world (next to not being able to continue to enjoy life with me) was that his grandchildren would not remember him as clearly as they would have had he lived until they were adults. Moreover, our son married in September 2006 and had no children by the time Don died in June 2008 but does plan to eventually have children. Therefore, Don knew that he would have grandchildren who would never know him.

"Don often said that his childhood was idyllic and thoroughly enjoyed reminiscing about his past, even though there were conflicts between Don's brothers at times, and Don tried to be the peacemaker. Despite the conflicts, Don loved his brothers and his parents dearly and had a good relationship with all of them."

Melissa describes the role of her dad in the family. "Life before cancer was much simpler. Dad (Jim) was a strong and distinguished, well-loved happy guy! He was quick with a joke and always had everyone around him laughing. When he walked in a room, he was like a breath of fresh air. He also conveyed himself in the colors he wore . . . bold and bright. Family gatherings were always centered around Dad. His strong personality governed the activities for the day. Family gatherings were always joyous and fun.

"My family was close knit. We knew what was going on in each other's lives at all times. I think Dad helped facilitate this. He was the glue of the family."

There are exceptions

Though many families describe the close-knit relationship they shared, there are others that don't follow this description. They may not only be separated by distance but also by what they have in common.

Such is the case for Fred and Lisa. As Lisa explains it, "Our backgrounds were as different as night and day. Fred came from Cleveland, Ohio. His family was a Polish Catholic—blue-collar, working-class family—with two sisters, one five years older and one five years younger. My family middle class, Jewish from Long Island, New York, and an only child. When we met at Syracuse University in 1972, he had already lost his father, and his visits back home were for holidays and sometimes vacations. I was very close to my parents and remember numerous vacations and many visits from them. When we met and fell in love, all the differences seem to bring us to a shared philosophy of life, values, and goals. Neither of us were the first choices of our parents for partners, but with time, they seemed to accept that we were a strong couple and made each other very happy. Our two children, Sari and Ryan, were raised with a sense of close family relationships, since we spent a lot of time with my mother (my father had died the year before Sari was born) and holidays with Fred's family. We also spent a lot of time together on family trips and outings."

Jennifer R. describes the closeness of her family despite the divorce. "My parents had been divorced since 1976, and both were healthy and superactive for their ages. I have always considered myself to be very lucky. I don't have a big family, but I always thought that I was lucky to have a small but close family. Family was a huge part of my life, and I considered myself blessed to have closeness in my life.

"My father, Brian, had moved to Colorado when my parents divorced in 1976 and remarried not once but twice. Since my father had moved away when I was so young (age ten), I only saw him two or maybe three times a year. My relationship with him was close at times but not like that that I had with my mother.

"Someone once told me that my mother, Ethel, was the best mom I could ever have. Since my father had moved away and was never around much when we were kids, my mom had done the bulk of raising us. My brother, Brian, and I were very close to my mother. We had our own unit—just the three of us—plus, our menagerie of animals that included three dogs, two cats, and at various times birds, fish, guinea pigs, rabbits, gerbils, and mice. We loved animals, and my mother gave us the chance to experience all the joys that they bring to your life and the commitment and work it takes to take care of them. Just one of the many lessons she gave us.

"My mother has always and will always be my primary influence in life, my friend, supporter, and wisest counselor. She had an enormous heart filled with passion and compassion and empathy. I like to think that I learned well by her teaching. My mother always said that I am my mother's daughter, which is the highest compliment I could ever imagine. We like the same foods, have similar tastes in so many things, and really enjoy each other's company. I always idolized her in many ways—her boundless energy, nonjudgmental way, kindness to others, and service to animals and humans in need. She was never afraid to get her hands dirty and get whatever job needed to get done completed. She rarely complained and had never been sick other than the occasional cold. Mom took wonderful care of herself, exercised, ate right, went for regular physicals, mammograms, and never smoked." (My father, however, had been a smoker for about forty years of his life.)

Sue explains the relationship with her mother. "My mother moved from California to Utah at my invitation in 2004. She had given Len and I a great deal of financial support for many years, and one of the expectations on both our parts was that we would take care of a lot of things for her, including home repairs and yard work. Len wasn't good at completing that type of task, but we all were hoping he would do that for her as a partial repayment of her generosity. She purchased a house for us in 2004, and six months later, I found a home for her a mile away. When Len's disability money came through that same year, we started paying her rent, which was a great relief for all of us."

Sometimes siblings grow closer as they grow older. As Mary explains, "When I was almost three, my family moved to a house in a small town in Iowa, where we would remain until all of the kids left home. My brother was the oldest, and I was the youngest. My brother got tired of the gaggle of four little sisters who invaded his privacy and dominated the telephone, television, and bathroom time. When he reached his teens, he moved a block away to stay in our grandparents' extra bedroom. He still came home a lot, however, and tried to boss his little sisters around. During those growing-up years, I mostly remember my brother as the guy who clonked me on the head when he thought I was making a face when Mom took pictures of us on the first day of school. Basically, we had the same love-hate relationship that most siblings have.

"Because he was more than nine years older than me, my brother was pretty much gone when I was around ten years old. He went away to college, and we saw him when he came home with his huge laundry bag. After he got married, moved to Pennsylvania, and started a family, we only saw each other a few times a year.

"When I was in college, my brother brought his family home for the Christmas holidays. Some family circumstances left the two of us alone at my parents' house one evening, and we started to really talk. We connected for the first time as adults and developed a much-deeper respect for each other. This bond was never broken, even after we both moved to different parts of the country."

Normal and uneventful

According to Beverly, "Our lives were very normal before the PC diagnosis. My husband was a college professor, highly regarded in his profession, with many scholarly publications to his credit. I thought he worked too much, but his work was very fulfilling."

According to Joe, "Life before cancer was like anyone else's. No cares, no worries, and above all, no concern about cancer. Cancer runs in my family. My father died of kidney cancer, and my brother has colon cancer, but I personally never worried about it.

"All and all, life was routine and boring."

Worried about cancer

Beverly writes, "My husband's family had a significant history of cancer. His mother and father both died in their mid-fifties from breast and prostate cancers, and his brother died a few days before his fiftieth birthday from pancreatic cancer. Therefore, my husband was concerned about living a healthy lifestyle. He never smoked, and we belonged to a gym. And although he wanted to lose a few extra pounds that had been gained over the years, all of his routine blood-work results were fine: he was a healthy individual.

"We were both grief stricken when Bill, my husband's brother, was diagnosed with pancreatic cancer early in 1988. At that time, we didn't have a full awareness of just how serious this diagnosis is. Amazingly, Bill (who lived on Long Island) was able to receive state-of-the-art chemotherapy locally, and his pancreatic cancer seemed to go into remission. We were very hopeful that he had beaten it, but sadly, after approximately a year and a half, Bill died, just a few days before his fiftieth birthday. We were devastated.

"As of the late 1980s, my husband had a familial connection to pancreatic cancer as well as the prior family history of breast and prostate cancer. However, in all of the years between my brother-in-law's untimely death in 1989 and my husband's diagnosis in December 2006, no physician ever mentioned that my husband was likely at an increased risk for pancreatic cancer. If we had known then what I know now, there are things he would have done, which might well have saved his life or, at the very least, given him many more years of a good quality of life!"

*　　*　　*

In thinking about our lives before pancreatic cancer, we remember what we enjoyed and what brought us happiness. The stories below illustrate the variety of ways that each of us found happiness, some very similar and others very different.

Having fun

As Bette tells it, "We worked hard, and we played hard. We had many toys throughout life such as boats, motor homes, four-wheelers, snowmobiles. Our boys motorcross-raced, and we all enjoyed camping, fishing, boating, and traveling. Our life had been a wonderful one full of love, laughter, and always excitement. There was never a dull moment, and he planned our activities to the fullest. Traveling was always a favorite of ours. Having been to Europe and many states have left memories for our family."

Camping with tents, a pop-up, and a motor home were Lisa and Fred's favorite activities. They loved to travel, cruise, and see the country. Lisa states, "We shared a love of photography, and our home is filled with videos, albums, and pictures on the walls of every vacation and place we visited."

Ryan talks about the memories he has of one of his favorite trips. "When I was younger, the four of us went on a cross-country trip. We had such a great time going to the Grand Canyon, Yellowstone, Mount Rushmore, and many more places. We had so many nights just around a campfire, roasting marshmallows, cooking on the grill, or just simply sitting outside, talking about where we were going to go or what we were going to do the next day. The one trip that I will never forget is when my dad took me snowboarding with the family. The reason I remember this trip the most is because my dad tried to go snowboarding with me. He wasn't the best at it, but he did give it his all and ended up getting a lot better toward the end of the day."

Traveling played a huge part in Pat's life. "Travel has always been part of our lives but especially since my retirement in 2004. Our first big purchase was a motor home, which was a very large part of our lives. We belong to a club with a great group of people, several of whom have become good friends. The club has an outing pretty much every month somewhere in California, and everyone was good fun. Other times, we took the motor home on trips to wherever we felt like going at the moment or combined a trip somewhere else with one of the outings. In 2007, we went to Mexico along with eleven other rigs, as we call them.

"We really enjoyed traveling other ways as well. We have been to all fifty states, mostly by car, spending enough time to be able to say there is not much of the United States we have not seen. When we finally stepped off a cruise ship in Alaska, Charles said "Fifty!" We also managed to visit about twenty-five countries, some independently, two on tours, and others as port of calls from cruise ships. We always took some sort of shore excursions and tried to see as much as we could. The best one ever was on the recent trip, when we left the ship in Port Said, Egypt, for a two-day tour, spending a night in Cairo. I would never have believed we could cram so much into two days, but we saw as much as humanly possible in that length of time. A fun highlight was an unplanned camel ride. I only wanted a picture, but the man who had the camel picked me up and put me on the camel. We have terrific pictures of the two of us on the camel, with the great pyramid at Giza in the background."

"We enjoyed walking, going to the gym, movies, being with family and friends, and (infrequent) vacations. In good weather, we also enjoyed day trips (an hour from our home) to walk on the boardwalk overlooking the Atlantic Ocean," says Beverly.

Carrie states, "My favorite free time was spent up at our lake cabin in northern Minnesota. We had ten years of hiking, fishing, four-wheeling, snowmobiling, and gathering around the bonfire making s'mores and wrapping up the day with full stomachs and happy hearts. Nothing made me happier than to wake up at the cabin to the smell of Mom's coffee and Dad reading the map to discuss which trail we were going to conquer that day. Even as adults, we kids brought our children along to begin instilling the love for the great outdoors at a young age, as our parents did with us. Other hobbies I had before cancer were running (ran several marathons), traveling, playing the piano, exploring back roads, and raising my two young children."

Melissa also enjoyed the outdoors. "My parents had a travel trailer that they kept at a campground near Bethany Beach, Delaware. Dad and Mom would spend weekends during the summer here. My brother's family and my family would visit an occasional weekend as well. Happy times were spent out on their boat. Dad would have us out, and he would head out to the ocean on the quest to hopefully see dolphins. We would sit there on the water, still and quiet, and eventually, they would appear. We would sit there mesmerized by their beauty and friendliness."

Deborah elaborated on what she enjoyed. "I always loved to spend the spare weekend at my parents' house, being taken care of and relaxing in their spacious house and watching the beautiful birds at the feeder in the backyard from the kitchen window, a stark difference from my tiny one-bedroom apartment overlooking another apartment. One thing that brought me great joy that I must have taken for granted, since I only realized this after my mom's illness, was being able to make my parents happy and proud of me."

Spending time with family

According to Jennifer R., "Happiness to me was spending time with my mom, shopping with her at discount stores which she loved, helping her with her cat rescue, and our weekly and infamous Sunday-night dinners. During the summer, we had dinner in our backyard. During the winter months, we had dinner at Mom and Jim's house. What I loved most was when we went out and did something together, got something to eat, and had really good talks. I loved our talks. I felt blessed to have such a wonderful mother that had remained so close.

"My husband, Brison, daughter, Eden, and our two dogs always bring me great happiness as well. Keeping busy with my animal rescue work, exercising, and trying to keep up with my life and job always kept me extremely busy."

Jenifer M. discusses her pride in her children. "What brought me the greatest happiness was seeing my children grow and realize their dreams. Feeling successful in raising my niece (with help) to become a straight-A student with manners and kindness. Happiness is seeing my children have kindness and compassion for others.

"I was an avid reader, watching very little television. Several times weekly we had meals at each other's houses. I just enjoyed life."

Joe adds, "In our free time, we saw the kids and grandchild. We traveled once a year for a week."

"Len and I both enjoyed doing genealogy and visiting with family members. His family has a ranch in Arizona, and we went there when we could—not every year but almost every year—to family reunions and to enjoy the ranch itself. Len loved it there. I didn't like it as much, being allergic and being a city girl! But I supported him in it, and the children liked it a lot when they were younger," explains Sue.

"I would say what brought me the most happiness was spending quality time with my mom and Pomeranian, Raider," says Sherri.

Kathleen states what makes her happy. "Vinny's love, companionship, sharing family events, being around and getting to know the children and grandchildren, planning a future together, sports, baseball, tennis, bike riding, the ocean—we lived in Long Beach—movies, gardening, music. He introduced me to opera, my work. I loved my work. I was a full-time teacher. In general, having the opportunity to share everyday life events with a man I loved and felt companionably with. We also had a similar level of energy."

Just being together

Aaron explains what brought him happiness. "I've always been happy with Frayda, and we always enjoyed each other's company. I found that a high degree of intelligence and a good sense of humor are the things that mean the most to me in a life partner. For that person to have good looks and be compassionate (and Frayda certainly had these attributes) helps as well, but the first two items are by far the most important. We spent a great deal of time together (to the exclusion of such things as my bowling with the guys), and never tired of each other, even though most of the time she could finish whatever sentence I started.

"I have no shortage of hobbies—reading, flying my airplane, photography, etc. Even though she always was a nervous passenger, she was willing to go places with me in the airplane. She had a good artistic background and took to photography as a duck takes to water. We built this hobby into a profession, winning many prizes, and exhibiting and selling in galleries. When traveling together, we didn't look for jewelry stores but, instead, looked for photo opportunities."

Jeanine agrees, "Being Don's wife is what brought me the most happiness. I hadn't dated many fellows by the time I graduated from high school, but the ones I had dated left a lot to be desired as life partners as far as I was concerned. Also, all the marriages I had seen were not anything I'd want to experience. In short, I was very disillusioned about males and marriage and had decided I'd most likely never get married. And then I met Don at the local community college. At first, I only participated in group activities with him and his friends and just thought of him as a cute little guy. After several months, he let it be known that he wanted to go on a date with me, just the two of us. After praying about whether I should encourage him or not, I agreed to a date. The better I knew him, the more I realized that he was one of those one-in-a-million guys, and my appreciation for him continued to grow during our forty-three years together. Just being with him and our children was pure joy for me."

"The time we spent as a family and still spend today is mostly about being together, enjoying good food and laughter. When we are not together, we often talk by phone just to share the day or week events," states Ernesto.

Mary states, "My brother traveled to my area a few times a year on business. We both enjoyed spending time together, so we always tried to plan these visits so we would have some extra time to talk and catch up with each other."

Religion and making a difference

Judy explains what an important role religion played in their lives. "Jim and I were both very involved in our church community since we were teenagers. We greatly enjoyed our time doing all of these things. We met in a church-sponsored youth group, and the various members of the youth group married each other. We are still close to people whom we have known since we were teens.

Jim was a Cub Scout and Boy Scout leader. We both sang and played guitar in our church choir for over thirty years. We were both involved in running and organizing church carnivals, our parish drama group, our MAC computer club, and many other interests. Jim and I were also involved as volunteers in organizing and running a youth basketball league for grammar-school boys and girls in our part of the city where we live. He was a mentor and a friend to many young people. We both mentored many young folks into adulthood. Many of the former grammar-school kids that we mentored have contacted us over the years in order to introduce us to their children."

Searching for happiness

Michele explains, "I was searching for happiness. My dog brought me happiness. I love her dearly. I had just met a wonderful man who became my boyfriend, and that really brought me happiness. I spent every day with him. He made me smile and laugh, and everything was wonderful.

"I enjoyed spending time with friends, specifically drinking wine on my balcony and talking about life and business ideas and complaining about bosses and boyfriends. I enjoyed cuddling with my boyfriend, sleeping in on the weekends with my puppy, Bella, pushing us off the bed. Going to the beach with the dog, watching her try to swim and find the ball that was twenty feet behind her was a good day. I didn't love my work situation, but I did love that I was good at it. I loved dressing up in a suit and making sales deals happen. I really liked developing business and meeting lots of people and socializing every day with potential customers. I enjoyed making money and supporting myself and being independent. I was successful."

Pursuing other passions

Cinzia elaborates on what makes her happy. "My life is not what one would consider a normal life. My work is multi-diverse, ranging from film producing and costuming to home building to creative strategic planning in business. Basically, I am an artist with an education in medicine. Most of my life is spent in the equestrian arena, participating in the sport of show jumping. I train seven days a week when I am not on the road traveling. This is my life, my sport, my passion, which I share with my partner and family member: a gorgeous Thoroughbred."

* * *

Before cancer, we all had a philosophy of life and living. Some of us thought about it and lived our lives based on it, while others never really spent time worrying about it. The following describes the similarities and differences in how we lived our lives before cancer entered into our families.

Living for today because you never know what tomorrow will bring

Men didn't have longevity in Fred and Lisa's lives. Fred's father died of mesothelioma at the age of fifty-two. Lisa's father died of a massive coronary at the age of fifty-seven. Fred was only seventeen, and Lisa was thirty. Both were affected by such a huge personal loss.

Lisa explains, "Fred and I came to the same conclusion—live life to the fullest. Live for today. Don't put off anything until tomorrow because you may not have a tomorrow. Therefore, we bought an RV in 1999 (we were in our forties), went on a number of cruises, made wonderful parties for every occasion, and had fun. I always felt that we did more than most people do in a shorter period of time because we didn't think we would have time. We turned out to be correct, and I don't regret a day all that we did, the money we spent, or the incredible times we had together."

Bette agrees, "Today, I live each day to the fullest and never hesitate to do things as tomorrow may be too late."

"My philosophy before cancer was to cherish the moments spent with the ones you care about and love. Each of those moments is a pure gift and is what makes the heart survive when there are only memories to go on. I love the motto 'Work hard, but play harder.' I always took the opportunity to go on an adventure, large or small—whether in your backyard or across the globe. Let the ones you care about know how you feel in case it's the last time you get to see or talk to them," explains Carrie.

Simple life philosophy

Aaron explains, "I guess that my main philosophy in life involves not sweating the small stuff, not crying over spilled milk, and not worrying too much about things that might happen, particularly when they are completely out of my control."

Ernesto adds, "Be generous, kind, and respectful, and love your family and friends as they are yours and with you forever."

A life of giving

"The remainder of my days is spent in human rights and animal advocacy issues. Whether it is trying to save a political prisoner's life or stop the needless slaughter of both land and sea creatures, I remain devoted to these causes. My family is involved in a philanthropic foundation that raises awareness of the lack of funding for arts education," states Cinzia.

Jennifer R. agrees, "Do unto others as you would have them do unto you is something I always strive to live by. My mom instilled the 'Inch by inch, life's a synch, yard by yard, life is hard' philosophy as well. Sometimes, life is hard, but I always knew that this too shall pass. As I grew older, I became more concerned with making life better for others. Again, I followed the example that my mom set for me, and unsaid acts of kindness were important to me as a way of life."

Sherri writes, "I've never had a life of extravagance but always had necessities, some things I wanted (not needed), and love. What hasn't been given to me, I've worked for, and I expect others should too. I believe you should be fair and polite to others, and I expect they will be the same with me. I hate not being greeted or spoken to in a drive-through line, as I believe it's their jobs to work with and subsequently greet the public, as stressing as that can be at times. I think you shouldn't say you're going to do something unless you intend on doing it. I think if people who work in the medical field have to be drug tested, that people receiving public assistance *absolutely* should be drug tested. Assistance is there for people who *need* it, not for people who are too lazy or are on drugs otherwise. I believe in being honest with people, no matter what. One of my biggest peeves is people not being honest with or loyal to me. Tell me something no matter how you think I'll react and then let me deal with how I choose to react. I don't understand the rationale behind these terrorist countries that all they want to do is fight and kill people. We have to live on this world with each other. We might as well work together to make it good for everybody. I'm not a prude by any means, but I don't understand the loose morals people live with nowadays. The world has become a place where people worry too much about themselves and greed and not enough about other people. I think even if I had an argument with somebody as a younger person, I've matured enough that I'd still be there for them if they needed me. If somebody called me in the middle of the night needing a ride (i.e., flat tire, etc.), I'd absolutely pick them up. After all, I'd want somebody to help me if I needed it. I think we can all have differences and still learn to live in this world together. If everybody was the same, how boring would that be! I think you're never too old to learn something new or do anything. I detest cancer for what it's done to me and my family (and others, of course), and I would be highly irrational and angry to learn that the pharmaceutical companies and testing facilities have developed a cure for cancer but would rather milk the system for money than to save lives (as is rumored). I think if you can help somebody truly in need (i.e., not a panhandler, who makes more money on the street than I do working a full-time job), you should."

Jennifer M. states, "My philosophy of life is to be kind to others. Treat people with respect and understanding. Give more than you receive, which is a much better way to live. Handle what you are handed, deal with it, and don't whine. Someone else has a more sorrowful story than you."

"The things that mattered most were people to love, work that you loved and felt mattered, self-respect, health, and hope. I believed in God, felt myself to be a spiritual person but was not hooked into weekly churchgoing. I had a sense of having been given a second chance with a man I loved, and I felt lucky," Kathleen explains.

An easy, carefree Life

Deborah explains, "My life was very self-indulgent. I don't consider myself an especially selfish person, but I was twenty-six, with no responsibilities, living in Manhattan, and enjoying my life to the fullest. I believe in the 'work hard, play hard' mentality, so I didn't feel guilty spending my money and time in ways that I could enjoy them."

Simply put, "Life was good. Life was happy. Worries were few and far between. We lived a full fulfilling life full of adventure, full of joy, full of hope" is Melissa's description of her life.

Religion is central in our lives

Jeanine explains the important role of religion in their lives. "Don and I became followers of Christ when we were children. Don grew up with parents who followed Christ and encouraged their children to do the same. My parents weren't followers, but I met Christ when I went to church with a friend and her family. We both have been influenced by Christ's teachings and spirit all our lives, before we met, and during our marriage. Therefore, our philosophy of life was based on the scriptural concepts of striving to love our Creator above all else and other people as much as ourselves."

"Our philosophy of life was/is that all things are connected, all of life has purpose, and there is a reason for everything. God is present in *all* of life, both the easy and the difficult, and we believe that we will be reunited one day, after I leave this earth to rejoin Jim in the next life," explains Judy.

Searching

"My philosophy of life was optimistic yet undetermined. I was searching for what makes people happy and what to strive for. I was successful at my career, but it was all about making sales and more money. I wanted more. I definitely believed in a higher power and reason for being here but not sure yet what it meant for me," Michele writes.

CHAPTER 2

Diagnosis

A doctor who cannot take a good history and a patient who cannot give one are in danger of giving and receiving bad treatment.

—Author Unknown

None of us are doctors, but most of us knew something was wrong. Some of us endured months of watching or experiencing unexplained pain, and others received the words with no forewarning that no one wants to hear—the big C. We all remember how we heard and where we were. That's because our lives changed forever in that instant.

Unexplained pain

Jenifer M. describes her symptoms. "I had been having a vague, nagging lower back pain for two to three months. I was seeing an arthritis doctor who had diagnosed polymyalgia rheumatica. I went along with that diagnosis, taking the low-dose steroids, but it never felt any better. I went to see my gynecologist at the recommendation of my friend and explained my symptoms of the vague backache. He checked me and said everything appeared good, but he wanted to draw some blood. He drew a CA-125 and a CA 19-9 (which is unusual to draw). He told me that the CA 19-9 was supposed to be less than thirty-four, and mine was 3,500, which could be indicative for pancreatic cancer. He called my primary care doctor, who told me to go straight to the hospital for a CT scan and he would meet me there. Within two hours, I was told that I had a 4.6 cm tumor on the tail of my pancreas."

Cinzia describes how she was told about her stepfather's diagnosis. "Knowing Stepdad had recent gastroenterology problems, we were told it could be something malignant. For two months, the doctor conducted tests. One day, as I was waiting for my dental appointment, my phone rang. It was my sister, asking if I had heard the news. 'What news?' I asked. 'Stanley has cancer,' she said."

Ernesto agrees with the same unexplained pain for his father. "The beginning of 2007 started out all right. Things were going pretty well for everyone. Then my father started to complain of stomach pain and indigestion. He went to see his doctor, who started to examine him and treat the symptoms. My father was never one to see his doctor on a regular basis. He did not have a yearly physical and did not take medication other than over-the-counter medicine. By late May to early June, my father started to experience some itchiness throughout his body, and his skin appeared to be darker than usual. In early July, my father underwent a colonoscopy as per the doctor's request. When these results came back negative, my father had a consult with the doctor, and it was during this consult that the doctor noticed my father's jaundiced appearance. His skin was yellowing, and the whites of his eyes were also beginning to yellow. The doctor ran additional tests.

"Friday morning, July 20, 2007, my sister's twentieth wedding anniversary, my father had an appointment with the doctor. The doctor had requested to see him as soon as possible. We sensed that it was going to be bad news. My parents arrived at the doctor's office at nine a.m., and I was on my way to meet them. I was traveling to Brooklyn from Long Island during the morning rush hour and finally arrived at nine fifteen a.m. I parked illegally because I was late. My mother was waiting outside the office for me. The doctor had already seen my parents and told them that the further testing came back positive for pancreatic cancer and the prognosis was very poor.

"I went inside and sat down in the waiting room next to my father, and the only thing he said to me as both he and I were trying to contain our emotions was 'Cosi e finito la storia,' which in Italian (my father's native language) means 'This is how the story ends.' The doctor then called us into this office and explained the findings to me and my parents in further detail. Unfortunately, the only thing we heard was that there isn't anything that can be done and we should get our affairs in order. I am certain that anyone reading this book understands exactly how that feels."

Carrie explains the significance of the day and the diagnosis. "The family always celebrates the Fourth of July holiday at our cabin in northern Minnesota. Along with it being my dad's (Floyd) birthday on the fourth, we have a wonderful time celebrating and spending time together. We went into the 2006 Fourth of July holiday a bit nervous because my dad had some issues with acid reflux and a gallbladder attack a couple of weeks prior (Father's Day to be exact). He went into the ER thinking he was having a heart attack or something, and they did some tests and found some spots on his liver. He then went to get a scan done, and they said it was nothing to worry about and things were fine. Well, my parents didn't feel quite right and decided to make an appointment at a leading hospital for the day after my dad's birthday, on the fifth of July. My dad was thinking nothing of it, and just before the fourth, he bought the dream motor home he and my mom had been researching and wanting to buy for a couple of years. He was ready to travel and see the country. For some reason, I think we were all holding our breath and really trying to savor these couple of days together because there were so many mixed comments from the doctors in the days prior that we knew something was going on with our dad. I remember the lump in my throat and the constant uneasiness of my stomach that whole weekend. The fear was there, and it wasn't going away. I decided to stay up at the cabin the day my mom and dad left to go to the appointment. I had my thirteen-month-old daughter with me, and we were going to stay and clean up from the weekend so that we could have fun, fun, fun until the last moment, and I didn't want to waste the time when everyone was there to clean. We tried to go to my dad's favorite place for dinner, but it was closed, so we went to a place we found open and celebrated one last time for dinner on everyone's way out of town. I was scared, and it was lonely back at the cabin, and I felt horrible. The next day, I was so very uneasy and couldn't concentrate to do anything. I took my daughter for a walk and tried to think positive. I remember thinking, 'He is at the appointment right now,' and I was trying to send good vibes. I couldn't walk, eat, clean, visit with people, or relax. I knew in my gut that something was terribly wrong. I went back to the cabin and played in the water with my daughter, and I heard the phone ring. I sprinted up to the house, and it was my mom. I heard the words that put me into shock, and I hardly remember anything after that moment. My mom said that they think my dad has pancreatic cancer and that kind of cancer isn't a good type to have. She said that they still have to get some results back but that it wasn't good. I remember my dad got on the phone and reassured me that he was okay, and just hearing his voice calmed me down. He said all the positive things he always does and that if it was this type of cancer, there were treatments and things they would do to try to fight it. He sounded calm and grounded. He told me he loved me and he was not going to panic because there was nothing he could do about what was happening. We needed to find more out and learn about what it was we were dealing with. Boy, were we clueless about pancreatic cancer when we started this journey . . ."

Jeanine describes how she and Don received the diagnosis. "Don initially felt pain in January 2006 and went to the emergency room at a government hospital. He was diagnosed with GERD and was sent home to treat that via over-the-counter medications. The pain continued, so he made an appointment with his primary care provider, and she thought it might be ulcers and changed the treatment toward that. This all happened over several weeks. By a Sunday in the middle of March 2006, Don's pain was unbearable, and he went to the emergency room again. This time he had a CT scan, which showed a mass on his pancreas. We were told that the radiologist would look at the image on Monday and someone would call us with his conclusions. Don started taking a prescription painkiller and was feeling normal on Monday. We were optimistically driving home Monday afternoon because Don was feeling good when our cell phone rang. It was the ER doctor, and he bluntly told Don that the radiologist believed it was advanced pancreatic cancer but wouldn't know for certain until he had a biopsy and examined the results. Don was told to call the next day and set up an appointment for the biopsy. Don was driving as we heard this news over the speaker on the cell phone. When the call was over, neither of us said anything for a long time and then Don said in a flat tone of voice, 'As soon as I heard the news, everything turned gray.' I think I said something like, 'They could be wrong. They won't know for sure until after the biopsy.' And then we drove in stunned silence the rest of the way home, only occasionally breaking the silence with comments unrelated to the diagnosis."

Lisa describes several months of unexplained pain. "After four months of intense stomach pain, multiple visits to doctor, and changes in medication, Fred and I ended up in the emergency room at a local hospital. By midnight, a resident came by and inquired whether Fred had had a CT scan. When he said no, the doctor ordered one with contrast. What seemed like an eternity of waiting, the doctor, looking grim, told us that Fred had a mass on his pancreas. We didn't know what that meant and didn't even know where the pancreas was or what it did. His blood work indicated a CA 19-9 of zero, so even the gastroenterology specialist was optimistic that this mass would probably be benign. There was no mention of pancreatic cancer until much later, after an endoscopic retrograde cholangiopancreatography (ERCP), and a diagnosis of pancreatic cancer written by an oncologist on the prescription. I looked up the word *adenocarcinoma* of the pancreas and found out that it was pancreatic cancer. I ran to the computer to research the disease."

Ryan states, "I got the phone call from my mom. I remember that I was at work. I didn't know what to say."

Misdiagnosis

Beverly explains the months before getting an accurate diagnosis. "My husband was in pain for two-and-a-half months before finally getting the accurate diagnosis. The first indication of a problem occurred at the end of September 2006, when he told me that he had pain in his left side, radiating from his stomach to his back, and I suggested that he see a doctor. Because it was a weekend and the regular primary care doctor was not available, we went to the local urgent care center. The first misdiagnosis was that my husband was suffering from an infection, and he was given an antibiotic to take for ten days. However, at the end of the ten days, the pain had not gone away, and he was seen by his regular primary care physician who ordered a different antibiotic to knock out the 'infection.'

"By the third week of October, a new symptom had occurred in addition to the ongoing pain—bloating in his stomach. We later learned that the fluid buildup that was causing the bloating was called ascites.

"When that second round of antibiotics didn't solve the problem and with the onset of the additional symptom of ascites, my husband visited a local gastroenterologist and began a series of tests: blood work, CT scan, and MRI. But our experience was different from other people for whom these scans provided a quicker diagnosis. The initial scans showed that everything was normal. We were very relieved that there was no cancer (or so we thought at the time). We later learned that the ascites may have obscured the pancreatic cancer, resulting in a scan that appeared to be normal.

"The gastroenterologist was focusing his attention primarily on my husband's liver because, apparently, liver disease is often associated with ascites, and heavy drinkers may develop this fluid buildup. However, my husband didn't drink liquor at all, so we knew that was not the cause of the ascites. My husband was also experiencing weight loss. The first GI doctor suggested having a consulting opinion with a second local specialist, who reviewed all of the previous scans and blood work, but he didn't diagnose the problem either.

"However, the second specialist recommended that my husband have a procedure called paracentesis to drain the ascites fluid. It was a relatively simple outpatient procedure in which a needle, guided by ultrasound equipment, is inserted into the abdomen and the fluid is removed. This procedure was done the day before Thanksgiving 2006. My husband felt much more comfortable after a large amount of fluid was removed from his abdomen. He had been taking Tylenol to control the stomach pain.

"In early December, when the local gastroenterologists still didn't know what was causing my husband's symptoms, we decided that it was time to go to New York City so he could be examined by a GI specialist affiliated with one of the major teaching hospitals. We brought all of the medical records that had accumulated during the past two months.

"The doctor reviewed everything carefully and did a thorough examination of my husband, with particular attention to his abdomen. One of the doctor's questions was 'What is the result of the CA 19-9 blood test?' We had never heard of that blood test, and although a substantial amount of testing had been done, neither of the NJ specialists had ordered the CA 19-9 blood test. The NY doctor said that the CA 19-9 is a tumor marker test for pancreatic cancer. This was the first time that pancreatic cancer had been mentioned as a likely diagnosis, and it was frightening to hear those words. The NY doctor was not certain that my husband had pancreatic cancer, but he was certain that the situation was serious and that my husband needed to have the CA 19-9 and also an endoscopic ultrasound (EUS), which would provide a definitive diagnosis. He also urged us to act quickly to obtain these additional test results.

"We left the New York doctor's office, upset that the situation was so serious and also angry because multiple visits to two NJ gastroenterologists had not yielded the significant information that we had just obtained.

"A day later, my husband had blood drawn for the CA 19-9 pancreatic cancer tumor marker test. As I recall, normal is approximately zero to thirty-five. My husband's level was more than twenty thousand! (Note: This test is not always an accurate predictor of pancreatic cancer. Some people who have this blood test do have pancreatic cancer even though their blood level is not significantly elevated on this test.)

"The almost-certain diagnosis of pancreatic cancer was provided by a surgeon at a New Jersey cancer center in mid-December. He had been given all of the test results, including the extremely high CA 19-9 level. The surgeon was very kind, and he tried to be gentle in giving my husband and me the devastating news. But we felt as though we had been hit by a truck when he said to my husband, 'Get your affairs in order. You probably have only six months to live.'"

Michele describes a similar case of misdiagnosis. "My dad was feeling really tired for a long time, and we attributed it to his hemochromatosis. His iron levels were very high, and in getting phlebotomies to get his iron levels down, they kept dropping and spiking in a matter that the doctor couldn't explain. They started to do more testing, and soon, he started getting jaundice. All this time, I was in San Diego and had no idea of the severity of the situation. I came up to San Jose for my dad's birthday, and he was totally yellow. I freaked out. He was also very irritable and not himself. I was terrified and vowed that I would not leave until we figured out what the heck was going on. He scheduled an MRI, and nothing showed up on that. Then he did an ERCP, and his gastro doctor came out of the procedure with a look that was undoubtedly not good. He explained to me that there was a mass that was blocking his bile duct. Although he didn't know for sure, it was most likely cancerous. My dad was 'out of it' after the procedure but understood the severity. The next day, we went to see a recommended surgeon and scheduled a Whipple for the next week. He had to go in for some blood tests the next day, and his sodium was very low. He wanted to put off going to the hospital, but I was convinced this was very serious and called the doctor. I explained that my dad had been having headaches, and I was really concerned about waiting until after the weekend. With the symptom of headaches, it changed the situation, and the doctor made him go into the ER that night. Apparently, low sodium can cause really bad brain damage, and he was concerned. We got to the ER, and they started a sodium drip on him. That was the beginning of a nineteen-day stay at the hospital. The doctor that admitted him had no tact or sensitivity and told my dad in a dubious voice, 'Get your affairs in order. You don't have much time.' This was before they were given a diagnosis, and this was a terrible experience for my dad. As they performed procedure after procedure to insert a biliary drain, they took a biopsy that came back positive. However, the surgeon was not convinced and wanted to get another opinion because the tumor showed no true mass on any scans. In any case, he wanted my dad to consider the CyberKnife radiation before the Whipple because it was a new technique that was working with good statistics for long-term survival, and they could also measure any potential metastasis if present and too small to see. We took his advice and went to a different hospital, and my dad got another biopsy. This was positive as well, and we had to accept the fact that it was cancer with grim statistics."

Sue describes a similar misdiagnosis. "In late October and throughout November of 2005, Len had stomach pains and difficulty sleeping. We went to a walk-in center and got a new sleeping pill which seemed to help a bit. We went to our family doctor, and he was given morphine pills and more sleeping pills. In early December, we went to a church Christmas dinner and party. He was very groggy, but the church was only a few doors away, and my mother was with us, and I encouraged Len to come. During the dinner, he became less alert, and his head almost fell into his dinner plate. Some friends helped walk him out to a friend's car, and we were encouraged to take him to the emergency room. One of my women friends went out to the car, thinking he was having a stroke but asked him if he had taken any medications before coming over to the church. (I had stayed in the hall with my aging mother.) She later told me he said he had taken a morphine pill and a sleeping pill.

"I went in another car to the emergency room, and we were there from nine p.m. until four a.m., at which time he said he wanted to go home. No one knew what to do or what the problem was.

"After seeing the family doctor again, who thought he might be having liver problems due to one of his depression medications, we scheduled a CT scan at the hospital. That was December 18, a Sunday. We went to church as usual and then went over to the hospital. We had our five-year-old granddaughter with us. The hospital was decorated for Christmas. Our family doctor had given Len his cell phone number and asked him to call as soon as the CT scan was read. Len did that. The doctor spoke with the radiologist and then told Len that he had a three-cm tumor on his pancreas.

"I don't think we got the full impact of what he said, but as we thought about it, we realized that what he was saying was that Len had pancreatic cancer."

Aaron explains his wife's ongoing gastrointestinal problems. "My wife, Frayda, developed some digestive discomfort and, after several weeks, went to a gastrointestinal physician for diagnosis and treatment. He ran an upper and lower GI series, didn't find a problem that would cause the discomfort, but gave her some medications to try.

"After several weeks of no improvement, the physician decided to probe further and conducted a CT scan. The result was that a large lesion, suspicious of carcinoma, was seen in the tail of her pancreas. In order to learn more, an MRI and PET scan were done, and the final confirmation took place after doing a biopsy, in which it was found that it was an adenocarcinoma. In addition, metastatic activity was noted in the liver and one adrenal gland."

Melissa explains her father's diagnosis, which was unclear. "It actually took about two months. We all hoped for the best results, but we all sort of feared the big C word. We were actually never told that it was definitely cancer until confirmation at the time of his Whipple surgery. The doctor made the determination that he had a tumor on the head of the pancreas but, without a biopsy, we could still hope it would be benign. After ten hours of surgery, the doctor delivered the news. He had a 6 cm cancerous mass at the head of the pancreas. The prognosis was not good."

Quick diagnosis

According to Judy, Jim made a doctor's appointment in early October 2005, without any urging from her. "I knew that something was bothering him. His doctor sent him for a gallbladder ultrasound. We were very concerned when that gallbladder ultrasound took almost forty-five minutes and the technicians kept asking us odd questions, like how much blood thinner did he take. They sent the results to his doctor, who ordered a CT scan. We still thought that the problem was his gallbladder.

"He had an appointment with his primary care doctor on October 18. He went by himself because we didn't think that anything was seriously wrong. His primary care doctor, a GI doctor, and an oncologist were at the appointment.

"I was driving home from work and was anxious for him to call me. I was afraid that he would have to have an operation to have his gallbladder removed, and I knew how much he hated hospital stays. By this time, he was slightly jaundiced.

"He finally called me on my cell phone and told me that the doctor thought that he had pancreatic cancer."

Bette explains the quick diagnosis of her husband's condition. "We had a CT scan and biopsy at a famous hospital and got the results the same day. Dr. Rubin came in and said he was sorry to say that Floyd had pancreatic cancer and metastases on the liver, and not just one but many. This meant it was inoperable and that there was only one option, and that was Gemzar for a period of time until it stopped working. He said there is only a 10 percent chance it would work."

Mary explains her brother's situation. "On his return from an overseas business trip, my brother, Gary, met his wife in New York, where she was attending a conference. Gary told her that he didn't feel well, and he rested more than usual in the hotel room. He asked her to make a doctor's appointment for him when she got home, and he went on to another business commitment. After he got home, he was seen and sent for an ERCP. While he was under sedation, the doctor passed a scope through the esophagus and stomach to the area in the duodenum outside the pancreatic duct. Dye was injected, and x-rays were taken of the pancreatic and bile ducts.

"After Gary woke up, he met with a surgeon, who explained that the ERCP had revealed a tumor in the head of the pancreas. The surgeon informed Gary that 90 percent of pancreatic cancer tumors are cancerous, but the surgeon was hopeful that this one was not cancerous. If it turned out to be cancerous, however, there was only about a 10 percent chance that the doctor could do anything that would cure the disease. If the surgeon was not able to remove the tumor, his prognosis was six to twelve months, but with chemotherapy and radiation, he might live around eighteen to twenty-four months. The surgeon had been trained at an institution that performed many pancreatic surgeries, and he was hopeful that he would be able to remove the tumor."

Kathleen elaborates on her husband's visit to New York City. "Initially, we were given the diagnosis over the phone by our primary care physician, who had requested a CT scan and had received the results. Three days later, we had an appointment in New York City and a repeat CT scan/contrast that confirmed the original diagnosis. This time, we were told face-to-face by the head surgeon who does the Whipple procedure. He spent less than ten minutes with us and told us the tumor was completely inoperable. He went over what they could offer us. He left the room to check the OR schedule because he said he wanted Vinny to have a stent put into his bile duct as soon as possible. He told us he would be back as soon as he did that. He never came back into the room. Rather, he sent his nurse in to go over the rest of the information with us. This was a terrible experience and completely unlike almost any of the other experiences we have had at this hospital in the ensuing year. Most of the personnel have treated us with gentleness and compassion.

"The surgical procedure the surgeon suggested was set for three days from the date we were told of the diagnosis. It was an outpatient procedure, but Vinny had complications a few hours after coming home, and I drove him into urgent care in the middle of the night. He was admitted to the hospital and stayed for a week with acute pancreatitis. The surgical procedure was called an ERCP. In that procedure, they placed a drain in the biliary duct so the pancreas can clear bile. They went with an endoscope and used that to place the drain. At the same time, they took a biopsy of the tumor through the endoscope to stage it. He had already been diagnosed with pancreatic cancer, but they did not know if it was adenocarcinoma or neurocarcinoma. Because there was so much manipulation of the pancreas, he developed pancreatitis as a side effect of the surgery, in addition to having cancer. The pancreas was inflamed and irritated because of all the manipulation to it during the surgery. He was in the hospital for a week while they treated the pancreatitis."

Jennifer R. describes quick diagnoses for both of her parents. "I knew my dad had pancreatic cancer prior to his surgery. After he turned jaundiced and after many doctors' appointments and tests, he had a stent put in to open the bile duct and reduce his jaundice. Prior to having the Whipple at the hospital, they greatly suspected that he had pancreatic cancer, so we were well aware of what it likely was. While my dad was having his Whipple surgery, his wife, my brother, and I waited in the waiting room. We got several updates during his eight-hour surgery that he was doing well. Upon completion of his surgery, the doctor came out and said he had a successful Whipple but that the cancer was large. They tested the lymph nodes, and several were cancerous. They estimated him to be at stage III–IV.

"With regard to my mother's diagnosis, we were told by all of my mother's doctors and after many appointments and tests—including CT scan, PET scan, several ERCPs, endoscopies, etc.—that there were abnormal cells in her bile duct area found on one of the endoscopies. None of the doctors ever mentioned pancreatic cancer. Each one was not even sure a full Whipple would be necessary but instead a shorter, easier version. However, from my experience with my father, I knew cancer was a possibility. I did not talk to my family about it, however, as I didn't want to scare everyone. When the doctor came out to talk to our family after the surgery, he told us they had found that she had pancreatic cancer. He said it was slow growing, and she had likely had it for a few years. She had clear margins, and no lymph-node involvement."

How can this be?

Jackie describes her visit to the doctor with her sister. "We went to the internist, my sister's regular doctor, who told her that she would like her to have a scan of the abdomen to see what is going on. We were then sent to a specialist for additional tests. He ran a CA 19-9 and reported to us that my sister had a 558. I said, 'What does that mean?' He replied that a normal test would be zero to fifty-five. I still had no understanding what this was all about. No one had said the word *cancer* yet. Then he told us what a CA 19-9 is and what a 558 could mean."

Joe explains his shocking diagnosis. "It started in November of 1999, when I went to my doctor for a routine blood pressure check. The truth is I was out of medication, and she wouldn't give me a refill without seeing me. She must have had a car payment due. As soon as she walked in the room, she said, 'Joe, tell me how you feel, and don't leave anything out.' I thought, 'God, I must really look bad.' I told her I had been having some stomach problems, but we had just come back from a trip, and because of a change of food and water, it messed up my pipes. She said I was jaundiced and wanted to know why. I said, 'Maybe I'm Asian. I read where one out of every three people in the world is Asian. Am I the third person you've seen today?' She told me that was not a good answer. I told her I was the accountant and she was the doctor, so it was her job to come up with a better answer. She said that by the end of the day, she'd have one, and she then sent me to the lab for a blood test.

"An hour or so later, we met back in the examining room, and I was informed I had a blockage somewhere in my system and was sent down for my first CT scan. After the CT scan, I went home. No one told me to stay. An hour or so later, she called and told me to come back. She said, 'Your gastroenterologist is waiting for you.' I didn't know I even had one of those.

"An hour or so later, I met with my new gastroenterologist and was informed that the CT scan showed the blockage was either a gallstone or a growth on my pancreas and the way to find out was to scope me, a procedure known as an ERCP. The good news was if it was a gallstone, he could possibly get it with the instruments at the end of the scope. I asked what if it was a growth on my pancreas. He said that if that was the case, they would do a Whipple. I asked what a Whipple was, and he explained that it was one of the most intrusive types of surgery there was. Twenty-four hours later, they did an ERCP on me, and when I woke up, they informed me that it was a growth on my pancreas and probably cancerous. I asked what now, and they said, 'Meet your surgeon.' Forty-eight hours later, they did a Whipple on me. I woke up the next day with a twelve-inch incision in my chest, four drain tubes coming out of me, and one feeding tube going in me, and an IV in my neck. I looked like a high school experiment that went bad. The pain was unimaginable."

Deborah explains her mother's shocking diagnosis. "I was out to dinner with a group of friends at a restaurant in New York City. My phone was on the table since I was expecting a call from my father regarding my mother, who had been in the hospital for a couple weeks due to a nasty ulcer. Her doctor had been going back and forth about surgically removing the ulcer. However, he ultimately decided to operate, informing us it would be a 'quick and easy' procedure. Sitting at the dinner table, I was mildly concerned since I knew my mother went in at five p.m., and it was well past nine p.m. before my phone rang.

"When he called, I picked up the phone and walked outside the restaurant. My father's voice was grave. He explained that during the surgery, the doctor saw that the ulcer could not be removed because the duodenum was fused to the pancreas. Looking further, he saw hard foreign chips in the liver. He immediately called the pathologist, who ran tests that showed the chips were malignant but did not originate from the liver. He told my father that they were most likely from the pancreas, but they needed to run additional tests to confirm this. The doctor was able to perform a gastrojejunostomy, which thankfully was a success."

Family, friends, and all our support systems had mixed reactions. People want to say the right things. Sometimes, there are no words. In addition, we were all trying to digest this shocking news and find the strength within us to adjust.

Shock and disbelief

Jeanine explains her reaction to the news. "At first I was stunned . . . in a state of shock. My mind couldn't process the doctor's words, and I was almost immobilized. Since Don didn't talk, I didn't talk. Later, when Don started crying, I also cried."

Sherri states, "I swear, for a split second after he told us what the CT scan showed, it was like I wasn't even there. My mind raced through what he had just said. 'Pancreas . . . mass . . . cancerous . . . oh my, pancreatic cancer . . . one of the worst kinds of cancer to have . . . this can't be happening.' My eyes started welling up with tears, and I'd glance over at Mom. She sat there, trying to be brave and strong. She gathered herself with tears welling up in her eyes and asked him exactly what that meant, and of course, we were told the worst. I was unable to hold my tears back and just started crying. The doctor stepped out of the room for a moment, and I got up, went to her, and we just hugged as tightly as we could, both crying. I felt as if the biggest horse you had ever seen had just kicked me in the gut with its two back legs (yes, it was that bad). It felt like I would either throw up or pass out."

Melissa's reaction was similar. "I remember returning to our hotel room with my mom and we were crying in disbelief. I did not want to lose my father. He was always there for me."

Lisa describes her reaction in three words: "Shock, desperation, and terror. It was like being put on a new planet. The whole world was different now."

Ryan describes his feelings. "I felt frustrated. I started asking questions, like 'Why me?' I was so angry."

Kathleen explains a similar reaction. "Going back to the phone conversation in which we were told and again in the surgeon's office, my initial reaction was heart pounding, followed by fear, then numbness. These feelings alternated. I remember I did not cry for about five days. While he was in the hospital as an inpatient, I remember traveling on the New York City buses and feeling a severe case of disorientation. I felt as if I were drifting out of my body. I was unable to speak words to people, and there were aspects that felt like a bit of a hallucination. That was semi-frightening, but there was also a quality of not caring if I were hit by a bus or car."

Shock turns to anger

Bette's reaction went from shock to anger. "A couple of days later, the anger set in to me. Why me, and why am I losing my soul mate? We had this perfect marriage of forty-six years, the kids were all grown, gone, and had families of their own, and we had just bought a new motor home to drive off into the sunset for retirement. It was what we waited for, for years. How could this happen to me? I got very angry but was immediately reminded that I had to enjoy the time I had left with Floyd."

Patients handle the news differently

Joe states, "I asked what a Whipple was, and he explained that it was one of the most intrusive types of surgery there was. I don't mind admitting I cried like a baby. I asked when he wanted to do the scoping, and he said tomorrow. I said, 'You're not wasting any time,' and he said, 'You don't have time to waste.' This is the first time I got really scared."

Cinzia describes her stepdad's attitude. "My stepdad instantly became angry and negative. He was convinced that treatment was futile and he was going to be dead in a very short time. His attitude initially was difficult, but I believed, with positive reinforcement, I could turn his head around. Boy, was I wrong!"

Jenifer M. explains how her shock turned to faith. "I was stunned, my friend fell apart, and my doctor (who is a friend of mine) was visibly upset. We walked over to his office, and he called the oncologist and made an appointment for me the next day.

"My only thought was how to tell my children, who were fifteen, seventeen, nineteen, and twenty-one. I called my ex-husband, my brother, and mother, and we all met at my mother's house. I told them, and we decided how to tell the kids. I never cried, but from that point on, for about two years, I felt surreal. My children all came home from their various schools immediately. We finally all cried, and I told them not to believe everything they read online. I told them not to get pregnant or married right away just so I would be there. I told them that I wanted them to continue with their life plans, which were to finish high school or college at that time.

"I distinctly remember sitting by myself two days after the diagnosis and saying, 'God, I am going to give this over to you. I have no control of the outcome, but I do have control over my attitude and will to live. So it is in your hands.' I was raised a Catholic and went to Catholic school from kindergarten through twelfth grade and think of myself as Catholic but kind of followed my own rules. Now that I really need God and all the saints, I have become more spiritual, if not more Catholic. Giving it over has released much of the burden placed on a cancer patient to overcome this diagnosis."

Michele describes her dad's reaction. "My dad was very optimistic and stayed that way throughout the radiation treatments. He would talk to his friends and tell them that he has the best doctor and we are doing the best new treatment. I was so impressed with how he handled it on the outside. Inside, I know he was really scared. We only really talked about things in depth a handful of times. I know he didn't sleep well, staying up thinking about dying. I tried to talk to him more, but he was so stressed out that making small talk or renting movies was a better solution on some days."

Pat talks about her husband's reaction. "In the early days, Charles was very accepting. Kind of a 'what is, is, I've had a good life' attitude. As soon as he recovered a little from his surgery, he became severely depressed, to the point of needing medication, which did help. He just mostly stopped talking about it, except when necessary due to practical matters. I, on the other hand, virtually never went thirty minutes without having his condition on my mind. Over time, that has changed for both of us, and now, over five months from initial diagnosis, it is all still there, but we are living somewhat more normally and involved in activities again."

Sue describes the change in Len. "Len was a very spiritual person. He had been suffering from depression for nearly twenty-five years at this point, but he kept his faith in God. In January, his older sister came up from another state, and while I was at work one day, they went to the temple where he heard Heavenly Father tell him that it was okay for him to go, that he had done what he could here on earth and he would be okay. He had a brilliant mind and had created inventions he felt he needed to bring to the world, but that day, he felt released from that responsibility.

"An amazing thing happened then. He let go of his depression. It wasn't instantaneous, but it was real. It was as if a huge burden had been lifted off his shoulders. He still had the pain. He was still resistant to receiving help from me and others close around, but another miracle occurred a couple of weeks later.

"I felt the need for a blessing from someone in the church. When asked what the blessing was to be for, I said, 'That Len would realize that what I am doing to help him I am doing out of love.' The blessing was given, and soon thereafter, he started receiving my help graciously and thankfully. I was very grateful for that change in his attitude."

"Fred's attitude was completely different," explained Lisa. "As the days passed, Fred seemed more in denial about the prognosis, and I became the research expert. I was determined to find out the best course of treatment and the options. This became my new job and responsibility. In the meantime, the rest of the family was in shock. However, Fred stated that he would be one of the 4 percent that would survive more than three years and refused to listen to any reports of doom and gloom. He also didn't want to hear about any research that anyone had done and refused to read anything about the disease."

"As time passed, I thought it wasn't going to be as bad as it really was because I didn't know anything about it," explains Ryan.

Jackie describes her sister's reaction so uncharacteristic of her. "I have never heard my sister say a foul word in her life, and I will never forget her turning to me with a look on her face that I've never seen and never will forget. With no expression whatsoever, she cursed. Later, outside the doctor's office and in the car, we talked a hundred miles per hour, and when I mentioned to her what she had said, she did not remember saying it. I mean, she said it right there and right in front of the doctor."

Bette explains a very different reaction from her husband. "My husband's request was to go create as many memories as we could with the time we had left. So in between the chemo and doctor's appointments, we traveled in our new motor home out East, went on a Panama Canal cruise with as many family members as could get off work to come with us. He felt good most of the time. We also attended a NASCAR race in Texas and went around Lake Michigan.

This was his request, to do the things he had wanted to do, and we did them all right away. He truly created memories for us. Every minute was enjoyed by family members. The closeness and time together were great. It was a time to share thoughts and to say we loved each other. We tried not to dwell on the thought he was sick. He asked us to enjoy him to the fullest while he was here, and we did."

Carrie describes her father's reaction and her feelings. "My dad said he would do his crying and grieving over his illness when he lay in bed at night. He was so strong for the ones he was worried about, made sure to talk about his faith in God, and was absolutely sure we would all be together again in heaven. I always felt close to my dad, and for me, I didn't have any 'catching up' to do because feeling close was constant throughout my life, and I have to say that was special to know it wasn't an illness that made me need to be close to my dad. We just were."

Reactions from family and friends

Cinzia describes the different reactions of her family. "I called my mother once my dental exam was completed. She was upset but said that another test (another EUS) was needed to confirm. I told her that I would begin investigating our treatment options. Mother was very upset, but the fact that I was on board seemed to lessen her fear. She knows I am very capable in the medical arena. Mother and Stanley had little experience in regards to dealing with serious medical conditions and were left treading water in a sea of sharks. My sister was sad for my mother and was wondering how she would cope with both the news and caring for someone with cancer."

Deborah talks about the difference in her brother's and her reactions. "My brother was a complete mess as well. Unlike me, he was in denial about my mother's diagnosis until the official results were out, about five days after I spoke to my father at the restaurant. Our reactions shifted. I freaked out the first few days and then was in my researching mode right afterwards. So when it hit him, I had already recovered a bit. My father, who has never been one to show his feelings, was amazingly strong. He was steady and supportive for my mother, who was still in the hospital, too weak from the surgery to start any type of treatment. My aunts, with whom my mom is very close, called almost every day and started their own information-gathering."

"We didn't tell anyone until after we got the biopsy results," states Jeanine. "When we started telling people, most tried to be optimistic for us. Some prayed and cried with us. Many didn't seem to know what to say, but they let us know they cared and supported us with their prayers. We had been volunteering in our grandson's third-grade classroom on a weekly basis before the diagnosis, and when the teacher heard what happened, she rallied the students together to make Don cards and give him gifts. It was very thoughtful and generous of the teacher, the students, and their parents . . . and we were deeply touched.

"We set up a CaringBridge account for Don, and many people consistently showed their love and support through their comments in the guestbook. That became one of the strongest sources of comfort and support for both of us."

"On the way home, I called family and friends and told them the news that Jim's doctor suspected that he had pancreatic cancer," Judy states. "They were saddened and very supportive. People kept saying that they would keep us in their thoughts and prayers. We always had a lot of support from our families and friends throughout our entire lives together."

Ernesto explains the different reactions of his sisters. "When we got back to my parents' house, I sat on the front stoop, waiting for my sister to come over. As she drove up, I couldn't contain my emotions very well and started to tear up and cry. She parked the car and walked into the gate, already crying, and said 'How bad is it?' I responded by saying, 'It's bad.' My father was in his room, the TV room. When my sister walked in, they both started wailing, and my father hugged her and said in Italian, 'I love you so much.'

"When I told my older sister what the diagnosis and prognosis were, she was also choked up with tears but was able to contain her emotions a bit more. My father loved all of us and still does from where he is now. But we all had a different and unique relationship with him. This was evident by the reactions we each shared with him that Friday afternoon."

Sherri describes telling the news to relatives. "After coming home from that dreaded doctor's appointment when we were told the CT scan results, I asked Mom if she wanted me to call my brother Jimmy in Alabama or if she wanted to call him, and she said the former. My nephew answered the phone, and I somberly asked to speak to his dad. I think even he knew something was wrong when I didn't talk to him first. Prefacing the conversation with my having bad news, I told him what we have all been through and the awful news we'd just gotten. I think guests he was having for dinner had just arrived, and he told me he was going into the other room so we could finish talking. He echoed my sentiment that he felt like he'd been kicked in the stomach and just could not believe this. (Bear in mind our father had just passed away only two months earlier. Mom and Dad were long divorced, but I was his caretaker too.) My other brother Gary, who lived locally, and I had gotten into an argument and weren't speaking at the time (we were soon afterwards), so I wasn't as concerned with calling him at that moment. (Mom would call him later herself to tell him the news.)

"Mom called her sister Donnie. She has two sisters, but she and Donnie were by far the closest, not only in age but with each other. She and Donnie spoke daily, so Donnie knew what had been going on and all the testing, etc. I later called Mom's other sister Katherine and broke the news to her. Their reactions were substantially the same. Neither could believe what was being told to them. They cried and said this just wasn't happening. (Also bear in mind that Mom had successfully fought and won her battle against colon cancer approximately eleven years earlier, and their younger brother had died about twenty years earlier from pancreatic cancer.) We had always been an extremely close family, so this was news they did not want to hear."

Sue describes her children's reaction. "Our daughter was living six hours away, but she came for a visit and to talk things over with us. She had a hard time believing that her dad was going to die. She eventually realized it, however. Our son, David, had just lost a close friend to pancreatitis. He was grieving over that quite a lot. He did a lot of research on the Internet and was shocked and discouraged by what he found. He lived forty-five miles away but came to help me out with some of the care, bringing his girlfriend, who was also very compassionate and a good caregiver. Sometimes, they would stay for the whole weekend to help. Friends from the church also helped with many things."

Jeanine describes the differences in her children's reactions. "Our daughter panicked at first. Since she's a biomedical research scientist, her first reaction was to do research. All the results connected to conventional medicine were dismal, but she found hopeful results in alternative/

complementary treatments, and she became enthusiastic about getting Don started on that track.

Our son, the eternal optimist, reacted by stating that he felt certain his father could beat this cancer even though the odds were slim. He became even more optimistic when he heard about the hopeful alternative treatments.

"Two of Don's brothers were supportive and caring throughout the duration of Don's illness. The third one said he was sorry this was happening to Don but had very little contact with him after that. That same brother also could not make it to Don's memorial service due to being 'unable to get off work.'"

Describing the differences between her mother's and father's situations, Jennifer R. explains, "When my father was diagnosed, his friends and other family members were very supportive. Everyone knew he had a mass that was likely pancreatic cancer.

"When my mother was diagnosed, my brother was silent and listened. My aunt and uncle were also silent and listened. Her husband, Jim (my stepfather), was completely taken aback as none of us thought this would be the case. We were all completely and utterly stunned, shocked, and in disbelief of what we were hearing. My stepfather made us all agree *not* to tell my mother that she had cancer. This would prove difficult over the next few days as, of course, she was asking the outcome of the surgery. Everyone kept saying we were waiting for the pathology tests."

Friends can play a critical role in changing an attitude. Mary explains, "One friend, who was a particular lifeline for me, made one statement that totally changed my outlook. She said that if it was my brother's time to go, my attempts to hold him here on earth could actually make it more painful for him. As soon as I heard that, I knew that I could not possibly do that to him for my own selfish reasons that I didn't want to lose him. Although I was still struggling to deal with the fact that he did not have a good prognosis, I became committed to doing everything that I was capable of in order to help him through what I knew were going to be some of the worst days of our lives. I would try as hard as I could to accept what was ahead."

The diagnosis sinks in

Aaron states, "After a few days, the initial shock morphed into numbness, but this quickly changed in that our entire immediate family—Frayda, our two boys, our daughter, and me—started searching for any treatment that might provide a better prognosis. When Frayda first was diagnosed with having pancreatic cancer, we hardly knew what the pancreas was, where it was located, and what its function was. We read as much as we could find in the literature and on the Internet until we felt that we were reasonably knowledgeable. Then we looked into experimental trials for which she would be a potential candidate."

Beverly explains what she and her husband did to take action. "When we were in the car, my husband and I had the same initial response: overwhelming fear, coupled with the attitude that we would find a treatment to prolong his life. We refused to accept the surgeon's view that my husband had only six months to live.

"Since we were both operating in the mode of 'we will beat this terrible disease,' my husband decided to be very selective about whom he informed of the diagnosis. We told only our immediate family members and closest friends.

"We both wanted to immerse ourselves in the most up-to-date information on treatments for pancreatic cancer, especially success stories. It was clear from the first Web site that I located that the prognosis for pancreatic cancer was very dismal, but we were determined to be as upbeat as possible. I reached out to a couple of physicians to ask if they knew of anyone who had been diagnosed with pancreatic cancer and was doing well. I was delighted when one of the physicians told me about a pediatrician in our community who had surgery for this cancer and was now doing very well. My husband called him the same evening that I obtained his home phone number and was very encouraged by the pediatrician's upbeat attitude and kindness in answering his many questions."

Lisa began to do research and started to call organizations for support. "As is my usual way, I will leave no stone unturned. Therefore, from the very beginning, I began to contact both Pancreatic Cancer Action Network and the Lustgarten Foundation for information and support. Pancreatic

Cancer Action Network had a support network, and I asked to speak with patients who had the same inoperable tumor. They gave me three contacts, who I immediately called. They were wonderful and honest in sharing every single aspect of their disease and their fight. I shared everything I had learned with Fred, although he wasn't interested in everyone else's stories. I found the stories reassuring and hopeful. Within three months, I had joined a caregivers group through Cancer*Care*. I also found a series of CD's from a book entitled *Fighting Cancer from Within* and bought them for Fred to use. I told him that I believed that the psychological fight was as important as the physical one. He listened to these CDs for the first six months."

Ryan's actions were very different. "I did nothing. I continued to live my life like nothing was wrong," he states.

"Mom wasn't what you'd call computer literate, so I did all the researching," explains Sherri. "Some things I told her, some things I didn't, but whether I told her or not, she knew it. In addition to research, we prayed a lot and we talked a lot and we spent so much time with each other. We decided right then that if we had something to say or felt like something needed to be said, we said it . . . no holding back, no regrets. At this point, if we held back, we wouldn't be hurting anybody but ourselves. We decided it was okay to be scared, it was okay to cry, and it was okay to scream if we wanted to. If we didn't want to cook or clean or even get out of our pajamas one day, we didn't . . . and it was all okay. Once you get this kind of news, *nothing* else is anywhere nearly as important as you once thought it was."

Melissa explains how her grief turned to action. "After a couple of days, I changed my grief into feeling like maybe I can help him find a cure for his disease. After all, he had a Whipple, so most of the cancer was removed. I could help him beat this.

"I sprung into action to be my dad's advocate. I researched everything I possibly could on the Internet. I joined chat groups, Yahoo! groups, etc."

Jennifer R. describes what she did for both of her parents at different times in her life. "I helped my father as much as I could to get ready to go back to Colorado. He would go back to begin his chemo/radiation once he was strong enough to travel. I thought he would be one of the lucky ones as he was so determined to beat the cancer.

"When my mother was diagnosed, I was eight months pregnant and determined that my mother was going to live to see her grandbaby grow up. I began doing lots of research in places that I had gone five years earlier. I could see that in five years, the Web information for pancreatic cancer had grown greatly. Pancreatic Cancer Action Network was a great source of information (much better than five years ago) for me as was the Lustgarten site. It was all I thought about day and night. I didn't care about my baby, my husband, my dogs, my other family, nothing. All I cared about was my mother. She was my best friend and confidante."

Jackie's family had mixed reactions going forward. "Once we were sure of the diagnosis, her daughter was like 'Oh, okay, so what do you do about it?' Her son asked no questions and just said, 'Wow.' Her husband isn't home very often because he is a long-distance trucker, and he doesn't do research on these things. Therefore, he really had no idea of what the diagnosis meant. Our other sister and our brother were the most affected by the news. We have all been very close and it was a 'Be strong' time for all of us. We all got busy looking for answers."

Deborah looked for resources. "I stumbled upon the ACOR pancreatic cancer LISTSERV around this time, and it was such a lifeline for me. I started to feel more positively about my mother's prognosis, and I was determined to get her away from our community hospital to a high-volume hospital. I did so much preparation with researching clinical trials, doctor, studies, etc. During this time, I found out about the Pancreatic Cancer Action Network and the Lustgarten Foundation. I'm an avid runner, and at this point, I thought about running a race to raise money for one of the charities. I convinced six of my close friends and cousins to agree to run a ninety-two mile relay as a fund-raiser for pancreatic cancer awareness. Looking back, I'm actually surprised at myself for planning this so early after my mother's diagnosis. My brother and I realized that both my father's father and my mother's mother had died from pancreatic cancer also.

"I went on a rampage to find out as much as I could about pancreatic cancer. Even though I respected my medical friend's opinion, I was sure there was another way. I was stricken and appalled to find out how little money is dedicated to pancreatic cancer and how little progress had been made. I thought back to Randy Pausch, the Carnegie Mellon professor who died from pancreatic cancer, and appreciated all that he had done to bring more awareness to this disease."

CHAPTER 3

Treatments

The greatest mistake in the treatment of diseases is that there are physicians for the body and physicians for the soul, although the two cannot be separated.

—Plato

As we each looked back on the decisions that we made and that some are still making about treatments, one word comes to mind: confusion. We want answers. But as we consult more doctors, we get different answers. Why? The answer is due to the fact that this disease isn't black and white. What works for some doesn't work for others. What are we left with? We are left with the important relationship with our doctors based on trust and the feeling that this person is looking out for our loved one or us.

Before we present our stories, it is important to understand the nature of this disease and the options (or lack thereof) that we had to face. Bev explains it best:

"One of the many problems that we discovered with chemotherapy treatment for stage-IV pancreatic cancer is that there are very few chemo drugs that have received FDA approval *for this particular cancer*. The drugs that did have FDA approval for combating pancreatic cancer were generally recognized as not being very effective. However, there are many drugs that have FDA approval for *other* cancers. When a chemo protocol is determined to be effective in stage III, clinical trials for a particular cancer, it then receives the FDA stamp of approval in treating that type of cancer. However, the clinical research conducted to find effective pancreatic cancer treatments has been woefully inadequate. It is generally recognized that many people with an advanced pancreatic cancer diagnosis (i.e., stage IV) will die in six to nine months, and most will die within two years of diagnosis. Yet many oncologists are unwilling to use chemo drugs that are FDA approved but not approved specifically for pancreatic cancer because clinical trials do not provide sufficient evidence-based data regarding their value to patients with pancreatic cancer. In addition, insurance companies refuse to cover many of these drugs not specifically approved for pancreatic cancer.

"Another significant problem is that many oncologists have had relatively little experience in treating pancreatic cancer. Sadly, unless an oncologist is affiliated with one of the major cancer centers, he/she may not have sufficient knowledge of state-of-the-art treatments or clinical trials that may yield better outcomes than the 'standard' administration of the two approved drugs, Gemzar and Tarceva."

The following section will provide information that describes the challenges the contributors in this book faced when choosing treatments for our loved ones. It also explains the very difficult decisions that we all had to make while trying to increase the longevity and quality of the patients' lives. Confusion is the operative word as each of us tried to find the treatment that would result in the best outcome. Here are our stories.

Making decisions

Ernesto explains, "When my father was diagnosed with pancreatic cancer on July 20, 2007, my family and I needed to make decisions regarding his care very quickly. Unfortunately, with pancreatic cancer, every moment counts because this disease is so aggressive. We needed to gather as much information about the disease as we possibly could, and we were starting from scratch because none of us was prepared for a diagnosis as the one we received on that day.

"The one benefit we had was that we could use the Internet to obtain a wealth of information. I, as well as my sisters and many other people that receive this grim news, spent many a night on various Web sites on the computer, trying to learn what we could about the disease—what the symptoms are, what treatment plans are available, and what the prognosis would be. Much of the information that we were obtaining portrayed the prognosis as bleak at best. The statistics surrounding pancreatic cancer are very scary. There is virtually almost no hope or promise for recovering from pancreatic cancer. Even living with this disease for any significant length of time is considered a rarity. Few people diagnosed with pancreatic cancer survive more than one year, and less than five percent survive more than five years. As I and my family witnessed firsthand, pancreatic cancer eats away at one's body very quickly. My father barely made it past six months before succumbing to pancreatic cancer on January 25, 2008. Because of these grim statistics, one of my sisters decided to leave the decision making to my parents and myself regarding all aspects of care and treatment for my father. Knowing what the prognosis was for my father, I felt so bad for my family and did not divulge all those statistics to my family members. With pancreatic cancer, I personally believe that quality of life is the single most important factor to consider. Since it is so difficult and rare to beat the disease, you should emphasize living your life as best you could, given the circumstances."

Ryan talks about the decisions that were made as a family. "We looked at all possible choices and went with the one that best suited this particular situation. The research I did personally was all on the Internet. I went to various Web sites, looking for anything that I felt could help. The discussions that I had with the family were pretty much all of our gathered research combined into what we felt were the proper steps to take at the time. The treatments were extremely tough at times. The reason for this was that I knew the extent of the disease, and I knew how painful it was just watching. The treatments were handled one step at a time. Some of the positives of the treatments were actually being able to see the tumor being shrunk in just a few months. Just to see that for some time, I thought that there was actually going to be some hope. Some of the downsides of the treatment were seeing my father start to get weaker and thinner by the day. Constantly seeing him in pain was extremely depressing because I knew that there was nothing that I could do about it. For the most part, I felt pretty positive about the treatments and tried to stay optimistic. It was hard to try and stay positive, but I had to because if I didn't, my dad would start losing faith in himself."

Pat explains the research they did to determine treatment options. "As soon as we learned about the tumor, we spent time on the Internet researching pancreatic cancer and found out about various treatments, including the Whipple surgery. The day after the biopsy, the surgeon, with whom we were very impressed, confirmed the diagnosis. He drew pictures for us on a pad of paper and explained everything he would do, including a vein graft. It appeared there was no artery involvement. He did tell us that he would examine the liver, and if there were obvious signs of cancer, he would not complete the Whipple but just close the incision.

"There was never any question in either of our minds about having the Whipple as we had seen the statistics that told us this was the most likely treatment to give Charles as many as five years with what was thought then to be a stage II or III adenocarcinoma.

"What the surgeon found that was not apparent on either the CT scan or the ultrasound was that the tumor filled the entire pancreas, was totally wrapped around an artery, and was pushing out toward the bowel, so it could not be removed. He removed the gallbladder, which was full of stones, isolated the common bile duct which contained a very large stone, isolated the pancreas, and connected the stomach to the small intestine. He told me that it would make Charles more comfortable and better able to eat. He also told me that night that we absolutely should consider chemo and maybe radiation. I asked "How long?" and he responded "Six to nine months, if chemo is effective.""

Michele explains how they reached out to someone outside the immediate family and the role of research. "We made decisions by having a close family friend come with us to doctors' appointments to make sure we had an objective onlooker asking questions we may have forgotten to ask because we were still in shock. We went to two oncologists and a gastroenterology surgeon. They all agreed on the treatment plan. We also trusted our doctors' judgment and took their advice on what was the best for my dad. We did get a second opinion on the chemo options down the line, but all in all, we trusted the doctors as they came highly recommended.

"Since the disease moves so quickly, you really have to trust that the people who are on your team will make the very best decision.

"I did a lot of online research but got very overwhelmed. There were not many good stories, and you can get very lost in reviewing the researching studies.

"I used online, Pancreatic Cancer Action Network, and the Stanford library. I called and developed relationships with the head physician's assistant at Stanford and UCSF and called them whenever I had a question, needed some reassurance, or had a complicated health question. They were wonderful, absolutely brilliant, and really helped me calm down and focus on the plan and try to enjoy our time together at the same time."

Melissa explains the role of her family and research. "We talked as a family about the best way to proceed. Most of the decisions were left up to my dad and what he wanted to endure. I researched so much online. I joined the pancreaticcancer.org group, which was invaluable to the family as my dad battled his illness. No matter what I would post, someone would answer. I don't know what I would have done without my computer and the ability to use the Internet. It was an absolute necessity for us to find out the latest studies and to also keep in touch with the doctors."

Kathy describes the people involved in making treatment decisions for her husband. "They were my husband, Vincent, my husband's youngest son Rob who at the time was about thirty-three years old, and I. We also spoke with another family member, a brother of one of our sons-in-law, who was a radiologist specializing in cancer work. At the time of the diagnosis, we physically went to two institutions, both major cancer centers in New York City. We also talked with physicians at major centers in both Baltimore and Dartmouth and sent my husband's scans and tests results there. We read several articles written by one of the main doctors we were considering in New York City. I took notes on everything, created a notebook for my questions and a file for answered questions. Just as an aside, that one notebook turned into five, each detailing different aspects of his treatment. There were discussions with various family members, and some gave their input, but the primary decisions were made by the original three family members. More opinions than that felt too cumbersome, and the three of us were the most educated about the situation, having done hours and hours of research. Ultimately, we decided to go with the doctor and treatment with which we felt the most comfortable. By this, I do not mean we were not afraid. I mean, we chose the doctor who was a top contender and had a compassion, humanity, and flexibility we felt comfortable within the midst of our fear. She was open to discussion, fully explained what she was doing and why, and was responsive to our needs."

Judy concurs with the importance of online research. "When Jim was told that he probably had pancreatic cancer, I went online and looked up pancreatic cancer and found that there wasn't much hope. I called Pancreatic Cancer Action Network and talked to a PALS associate, and they sent me their booklets about pancreatic cancer, clinical trials, and nutrition. I also used the Internet, Johns Hopkins's pancreas cancer discussion board, the ACOR pancreas cancer LISTSERV, and the nurse at the surgeon's office."

Sherri describes the role her mother had in her own treatments. "My mother, Margie, made all of her own decisions throughout this whole process. I guess you would say I was the only (or main) one she discussed them with initially, and she sought my opinion, but I told her it was ultimately her choice. She listened to the various course of treatments the doctors told her about and to the information I gathered on the subject. She decided to go with chemotherapy (she was stage IV, inoperable) to fight this horrible beast. I had discussed with her the choice of getting a second opinion, but she did not want to have to deal with that, saying we knew it was cancer and that they really weren't going to tell her anything any different from what she has been told already. She stated that she didn't want to have to go through all that testing again. I respected her decision, and we moved forward without discussing it further. Other than listening to what all of the doctors told us, I immediately hit the Internet. Sometimes, I found, it's not good to read too much about certain subjects because you get all sorts of information that starts jumbling around in your head. It can confuse you, and you will probably get more upset with the more information you have once you realize how cruel this disease will be. Trying to keep that in perspective, I did read up on as much information as I could online. You begin to see that a lot of what you read will be repeated throughout numerous websites. It's just usually worded differently. Additionally, I had a friend/coworker who was battling cancer at the time (who has, sadly, since passed away) who would share so much information with me, the kind of information you won't read about."

Sue explains the role of their primary care physician and a close friend. "Our primary care physician referred us to specialists who would determine whether or not surgery was possible. Part of this process was to be marked for radiation sites. Len was enthusiastic about this, and as a person with a scientific mind, he enjoyed the procedure and chatted with the technicians who performed the procedures. One of the specialists we went to performed a bile stent. In a few months, it was clogged up, and I think he had another procedure to insert a stainless steel one instead of the plastic one that he had originally.

"We went to Salt Lake City for a biopsy, which showed the cancer had spread to the liver. I believe the results of this test narrowed our choices as it was determined that Len was not eligible for radiation or surgery. We were referred to a cancer specialist, and she presented us with a plan for chemotherapy. After she explained her plans, we asked her to leave us alone so we could talk and pray. We decided to go ahead with the chemo.

"One close friend we hadn't seen for a number of years e-mailed us an explanation of a protocol that didn't involve medicine, and another friend recommended XanGo. Len never had much interest in natural healing, and although I did, I felt this decision was more his to make than mine.

"The first thing I did was to look up pancreatic cancer on the Internet. We didn't know a thing about it. Our oncologist gave us some printed materials containing quite a bit of information. We studied it. We started attending a cancer support group once a week in the next town. If I remember correctly, someone in the group recommended a support group for pancreatic cancer patients that we could join online. Len was in too much pain to do much about it himself, but I joined the group and studied a lot.

"We got hold of a printed book with stories of pancreatic cancer survivors. The theme in that book seemed to be that fighting the disease was a moral responsibility and that one should search for the best care possible. We had very limited financial resources, as Len already was receiving Social Security disability benefits for the depression and anxiety.

Some success stories were shared through the Internet support group, but we kind of felt that their greatest push was to find the best pancreatic cancer doctor. The consensus seemed to be to find a doctor with a lot of experience with pancreatic cancer, and the closest one I could find was in another state. Len didn't really feel up to making that kind of a trip since our residence was in Utah."

"Deciding between alternative and traditional treatments was the issue facing my family," states Jeanine. "Our family researched treatments for pancreatic cancer via the Internet and discussed them with alternative health practitioners. During the whole time, we were praying for divine guidance and then followed the course that seemed to be the most promising.

"The research we did was via the Internet, books, and alternative health practitioners. We did get a second opinion from an oncologist at another educational health institution before Don started radiation and chemotherapy. He suggested that Don begin an aggressive chemotherapy regimen after he completed the current radiation and chemotherapy treatments, one that the other facility wouldn't approve even if Don had wanted to go that route."

The role of researcher was solely up to Cinzia. "Treatment decisions were made based on research available from doctors, Internet, friends, etc. As soon as my dad was suspected of having a GI malignancy, I began researching and speaking with leading cancer centers, nurse educators, and perusing scientific articles on the Internet.

"Once he was definitively diagnosed, the local oncologists were saying zero to three-month survival. We had friends at a local hospital that made an immediate appointment for a consult with the head of Pancreatic Cancer Oncology. We also knew that a well-known doctor at a local hospital was currently treating a family friend with stage IV metastatic disease, same as Dad, and she was doing well at the three-year mark. She had also been given the zero to three-month survival rate by a doctor in another city in the state. Additionally, we submitted records to another highly recognized facility (Dad's second residence was in another state) and considered an additional facility with a good reputation in a third state. In the end, Dad agreed to have a consult at the local hospital, which I felt did not go well. The next stop was going to a highly reputable oncologist known for treating pancreatic cancer, and that is whom we ultimately chose for treatment.

"Basically, it was up to me to present my mother and father with the best possible scenario. One of my mother's close friends on the staff at the local hospital made the call to Oncology for an immediate appointment. My dad did not want to travel nor did he want a second opinion. However, based on the amount of errors that had been made on his case in the local area and the lack of interest the other physician had in my father's case, I felt it was prudent to push for a second opinion at the other facility. My dad was very tense and agitated about his diagnosis. My mother and I finally convinced him (after he canceled his first appointment) to travel to the large, highly reputable treatment facility. In spite of the heated discussions, this was the best decision we could have made. I would advise all patients and caregivers to get a second opinion regardless of how much effort it takes to mobilize the troops."

From a patient's perspective, Jennifer explains the role of her primary physician. "My primary doctor was a great help in making decisions for treatment. He referred me to a local oncologist. She didn't impress me too much on the first visit. But I also had an appointment at another facility. I saw a surgeon who felt that I would be operable after a course of chemo. I saw the oncologist in another city on the same day. He recommended a phase 1 clinical trial. This all happened within days of diagnosis, and I was still numb. I signed up for the trial after no research. Knowing what I know now, three years later, I know that this was a last-ditch effort and I was just beginning, so a phase 1 trial was not the right choice. I began treatment the day after the appointment. Two days into treatment, I was in the hospital for five days with chemo-induced meningitis. I came home and decided I wanted to see the local oncologist. I saw her again and was very happy with her plan of care, which was one of the standards of care—Gemzar, Taxotere, and XELODA.

After beginning treatment locally, I did some research online. My friend did most of the research because I really was not mentally capable of it. I used Internet, referrals from doctors, and Pancreatic Cancer Action Network.

"My family and I had many discussions all during this time. Everyone wanted to know what the plan was. I'm sure my children did their own research even though I told them not to believe everything they heard. My children and friends went to all doctors' appointments with me, and everyone agreed with whatever I decided to do."

Beverly explains the issue of travel in their final treatment decisions. "We discussed the pros and cons of traveling from our home in central New Jersey into either New York City or Philadelphia for treatment at one of the major teaching hospitals in those cities. But when I researched the success rate of the Gemzar, Taxotere, and XELODA (GTX), which was developed by a nationally renowned oncologist at a hospital in New York City, I discovered that my husband's oncologist used the same protocol. I realized that my husband could get a state-of-the-art experimental treatment much closer to home and with a very caring and compassionate oncologist. We made the decision quickly for him to be treated locally, and we never regretted our decision."

Jennifer describes similar situations for both of her parents. Her father followed the advice of his doctors, but in 2001–2002, they were using the same chemotherapy they had been using since the 1950s. He (her father) was convinced that a clinical trial was the way to go.

With regard to her mother, "We made decisions as a family and followed much of what the doctor suggested. In the end, in August 2008, we found out the cancer had returned after having a clean scan only six weeks prior. She decided not to pursue any additional treatment. Her pain continued and only got worse.

"The research we used for my father was different since it was now 2001. The only really good resource out there was Pancreatic Cancer Action Network. The statistics were bleak, and he knew it, as did I, from reading about pancreatic cancer on the Internet.

"When my mother was diagnosed, we contacted Pancreatic Cancer Action Network so that she could talk to a survivor and someone who was close to her age. I was always looking for anything and everything I could read on pancreatic cancer. I still have a Google search for pancreatic cancer set up so that each day I get at least fifteen links with information from articles, research, etc.

"During my father's illness, my brother and I talked to my father and his wife about treatment. During my mother's illness, we had many discussions with family and friends."

Differing opinions

Deborah describes her feelings about the treatment decisions her parents made. "My parents were surprisingly unaggressive about treatment decisions. I think they were still in shock from the suddenness of the diagnosis and just wanted to be told exactly what to do. I was a strong proponent for having my mother treated at a hospital that specializes in pancreatic cancer rather than a local oncologist. I feel like I did most of the research into treatment regimens, doctors, hospitals, and alternative medicines. Though we may have felt otherwise, my brother and I agreed that the choices should be 100 percent my parents' decision.

"I did most of my research on the Internet. I combed through PubMed, signed up for Pancreatica, Lustgarten, and Pancreatic Cancer Action Network LISTSERVs, and joined the ACOR mailing list. I found there was a huge dichotomy between the published numbers and people's personal experiences. It was just so shocking to find out what a devastating disease pancreatic cancer is, but at the same time, I was so hopeful after hearing of several ACOR members' personal experiences at best-of-breed hospitals.

"My brother, father, and I spoke a lot about treatment options. My brother and I were very strongly supportive of my mother being treated at a large and more well-known treatment facility. My father was more hesitant due to the logistical problems. He has nontrivial health problems of his own, and due to severe lower back pain; he was not comfortable with my mother being treated far away from home. I felt the most strongly about this, though, and I offered to relocate to another state if it meant she would go to a better-known treatment facility. After being seen by both a local oncologist and one at a better-known treatment facility, my parents ultimately decided on the local oncologist. My mother was simply not in the condition to be able to travel or relocate. In my opinion, she was never able to totally recover from the ulcer surgery. In any case, we were comfortable with this decision because the out-of-town doctor sincerely endorsed the local doctor and assured us he was an excellent physician."

Bette describes the confusion of research. "We did a lot of research on the Internet, and with all the tools and information we gathered, we became confused as to what to do. Quality of life was an important factor for my husband. He did not want to suffer and go through a lot of torture to be alive. He wanted quality even though his life possibly would be shortened. He researched all the places in the world to get treatment but chose to stay at a well-known facility, which had a good resource center. At one point, we considered a treatment in Mexico, but the decision was vetoed quickly.

"Also the Internet was helpful. It came down to the bottom line of no matter what decision we made, there were few successful stories of people with pancreatic cancer. We also obtained a lot of information from the American Cancer Society. We did not know about Pancreatic Cancer Action Network at the time and wished we would have known about them for support.

"We, as a family, met often to discuss the situation and also had family members come to doctors' appointments with us. It helped to stay close, so all had a good understanding as to how this disease spread quickly."

From a daughter's perspective, Carrie describes the treatment decisions. "My father made the choice of the doctor from a large, well-known treatment facility. From the first visit, he felt very comfortable with his doctor and trusted him completely. After the diagnosis, we did a lot of research on the Internet, which was very depressing and scary. We also used the clinic as a resource for answering questions. We did discuss everything about treatments with our immediate family, and when we could, my brothers and I would go to appointments with my parents at the clinic. The final decisions were always made by my parents, but they always shared and asked our opinions regarding everything. It made our family even closer. It did take me longer to understand why the Whipple procedure couldn't be done on my dad, but the surgeons finally got it across to me that putting my dad through it would not change the outcome of his cancer. They see it day after day.

"Because my dad was feeling well and showed no signs or symptoms, we immediately went into denial. He was told he should start chemo right away. He made an agreement with his doctor that we were going to go on a two-week trip out East in the new motor home my parents has just bought the week before his diagnosis and would stop for the first treatment on our way home. My dad handled treatments well and was rarely sick. He was a tough Finlander, and his chemo did shrink the tumors and give him two extra months with us that were pain and almost symptom free."

"Making decisions was very confusing," explains Lisa. "First, we saw an oncologist because we had been told that the tumor was malignant, even after having been given hope since his CA 19-9 was zero. Then the oncologist asked us why we didn't see a surgeon first. No one really explained to us the order of anything. The most upsetting thing for Fred was that he was always in pain and needed painkillers just to get through the day. I did all the research. I immediately went onto the Internet, found Pancreatic Cancer Action Network, and called them.

"Pancreatic Cancer Action Network gave me three people to contact, who had the same diagnosis as Fred. I called all three and felt very reassured as they explained the course of treatments and said that they were doing pretty well. One person I called was actually on a commercial for one of the major cancer treatment centers, and she is known as somewhat of a miracle. I felt very hopeful. Then I went on a medical Web site, and the prognosis was rather grim. The more I read, the more depressed I got.

"Fred and I spoke to the kids and explained that we would do everything we could to increase Fred's chances for long-term survival. We also explained that we would fight this disease and that Fred was not a quitter."

Experimental options and complications

Aaron describes the treatments that his wife began. "When Frayda's pancreatic cancer first was uncovered, it already had progressed to stage IV, with extensive metastases. The prognosis was rather dismal, with an expectation of survival of only three months.

"We (Frayda, our three children, and me) uncovered an experimental program after visits to cancer treatment centers in three major hospitals in Boston and Miami and research on the Internet. We were able to enroll her in a program administered by an oncology medical group in another state. The program involved one several-hour infusion each week, with a one-week pause after every three weeks. Frayda felt fairly well during these 'chemo vacations,' and we found we were able to take some trips during these times.

"To ease the administration of the chemo, a 'port' was installed above her chest, next to one of her shoulders. Although this normally is a fairly routine procedure, it created a situation that was life threatening and sent her to intensive care for several days. During the procedure, an artery was cut, and by the time it could be repaired, more than a quart of blood had leaked into her abdominal cavity. A new port had be installed near her other shoulder (because of damage to the area where the first attempt was made), and the blood had to be drained. A considerable amount of blood remained, but this was left to be absorbed by her body. The whole affair left her tired, but in spite of this, the chemo was started."

Relationship with doctors

Some of us had excellent relationships with our doctors, and others never really felt that the doctors were 100 percent on our side. Here's what we went through.

Bev states, "Our excellent relationship with the oncologist, as well as the nurses, was a very important contributor to my husband's positive attitude. We had full confidence in the doctor's knowledge of pancreatic cancer, including how to deal with the various side effects from the chemotherapy. Furthermore, the oncologist was very reachable whenever we had concerns about the cancer treatment, returning our phone calls promptly in a calm and reassuring manner. As a patient at this facility, my husband also had access to a very helpful team approach to cancer treatment. On our first visit to the cancer center, we received a booklet describing all of the 'extras' that were offered *at no extra charge*—music therapy (from a very kind and caring music therapist), nutritional counseling (which my husband received at almost every chemotherapy appointment), innovative therapy groups, individual counseling, and lectures on various topics of interest to cancer patients and their families. There was also a free support group specifically for pancreatic cancer patients and their family members. Many hospitals offer support groups for cancer patients and their loved ones, but it is rare to have a support group geared specifically to *pancreatic* cancer."

"Our doctor at the facility and his assistant were great," states Bette. "They took time to explain and to listen. They at times felt helpless but did search for treatments with some help. Unfortunately, by the time we found out about this disease, it was so far gone that there was not much that we could do to provide him with the quality of life he wanted. On our last visit, there were tears in the practitioners' eyes. She had gotten to know our family so well and said my husband was the strongest and most positive person she had dealt with. They had a bond that was special."

Jenifer explains her relationship with her doctors. "My relationship with the doctors has been superb. All the doctors communicate, and that is very helpful. They are all happy with my progress and continued good health."

Cinzia had a mixed experience with the doctors. "The relationship with the local oncologist was very positive for both Mom and myself. The initial consultation revealed very quickly that my dad would be treated by a doctor who was extremely thorough, was clearly an expert in the field of pancreatic cancer, personally treated each patient without partners or what I refer to as the chemo mill, was concerned, honest in relaying what the odds were in a very positive manner, stressing that there was indeed hope. It was clear to me that in doing his job, he was there to save my dad's life and beat the cancer. Bravo! My dad was not accustomed to having someone tell him what to do. This created a rift between them. I will say that my dad did not follow any of the recommendations of his oncologist. I say this, knowing that there is a gray area of the unspoken rules during cancer treatment. A patient who wants to survive gets on board and sticks to the plan. My dad rebelled, creating a very frustrating situation for the doctor who was trying to save his life.

"Dad was much happier with the northern community doctors, who actually created severe pain and illness. Frankly, they wrote him off the day they saw pancreatic cancer on his record. Therefore, they asked him for nothing nor did they possess the passion or 'fight for survival' mentality of the oncologist from the other hospital.

"My relationship with his community oncologist was frightful. They were beyond incompetent and refused to admit error, even when they diagnosed a bowel obstruction as nausea from chemotherapy and allowed him to vomit feces for seven days. I had to travel to another region, get him to the emergency room, and fight with the oncologist to admit they had missed a bowel obstruction when it was clear from the recent CT scan that the tumor was beginning to obstruct."

Sue agrees with the mixed feelings about the doctors. "We didn't care for the oncologist in charge of our case, but there were several other doctors who were compassionate and with whom we related more positively. Len was a very friendly and bright guy and enjoyed conversing, so he developed a personal relationship with a few of the doctors."

Aaron states, "I feel that the doctors did their best during the course of Frayda's treatments. They were careful in the administration of her drugs, kept good records, handled her cancer-fighting drug regimen well, and were quite willing to spend time with us to explain their view of what was going on and address our concerns."

Another positive relationship is described by Sherri. "Mom was never one to not get along with anybody if it could be helped. We both, I would say, had a positive relationship with all of her doctors. I appreciated the way they would make her feel so special, like she was the only patient they were seeing that day, and I'm sure she did too. Obviously, we didn't like getting bad news from the doctors (so that kind of cast a negative light), but we both knew and remembered that this wasn't their fault, and we appreciated the honesty and compassion they provided, but we did have our favorites. We were extremely fortunate in this regard to work with such professional individuals. We realized, specifically with the oncology office, that they dealt with this kind of horrible news day in and day out and lost many patients to this disease that they probably came to know very well. It amazed us that they were able to continue showing compassion to us and others."

Deborah explains the disappointment she felt with her mother's doctors. "Her general surgeon at the medical center performed her emergency surgery very ably, but he basically told my brother, 'If I were your mom, I'd play a lot of golf and drink a lot of bourbon for the next few months,' implying that he had no hope for her recovery. Her general doctor at the hospital found a 'blockage' in her system and planned a surgery to remove it, before being reminded by my father that the blockage was actually the result of the emergency bypass performed during the ulcer operation. It is inexcusable that her main doctor did not care to look at her medical records and almost had her in for a surgery that was completely unnecessary and would have further compromised her already weakened state.

"Her oncologist never gave us much hope. He would answer questions fully but never volunteered any information. And as I said earlier, my parents were very passive, so they would not ask for his reasoning behind ordering certain procedures or treatments. This was very frustrating. I didn't meet the other oncologist that my mother met, but my mom was very impressed by him. Unfortunately, as I mentioned earlier, my parents decided not to be treated at another leading treatment facility."

Lisa explains her disappointment with the doctors. "I came with my list of questions. Not all the doctors were receptive to them. I explained that we were very informed patients and that the more we knew, the better off we would be. I'm not sure they all felt that way. When the first oncologist left for an unexplained reason, we had to ask for a replacement that specialized in this kind of cancer. It took awhile for this to happen, and in the meantime, we felt abandoned. When the first line of treatment was completed, they offered us no additional hope for second-line treatment. The second oncologist we saw was very honest with us but didn't seem as aggressive as I thought he should have been, especially when Fred began to develop other symptoms."

Jeanine describes a similar situation. "We didn't like the government oncologist because he was negative and discouraging. The radiation oncologist was much more encouraging and hopeful, so we were thankful for him. The palliative care doctor was kind and helpful but was always trying to make us 'be realistic' and not put so much hope into the alternative treatments. The integrative medicine doctors and other alternative treatment practitioners were cautiously hopeful but never promised a cure. They were realistic about the deadly power of pancreatic cancer and were focused on trying to provide Don with a good quality life for as long as possible."

More frustrating relationships are described by Pat. "Interaction with the doctors in one town was absolutely the most exasperating experience imaginable. There was only one we truly didn't like, but there were just so many of them, and they never all explained their roles in Charles's care. The surgeon was wonderful, but we didn't see a whole lot of him after the surgery. He stopped by to chat a couple of times but really was not involved in subsequent care. There were so many doctors and physician's assistants that I never learned all of their names. The most frustrating problem was that *no one was in charge.* When I was at my wit's end with receiving three or four conflicting pieces of information in a few hours, I calmly and rationally talked with one of the outstanding nurses about it. I'll never know how I did that because I certainly didn't feel calm! She contacted one of the doctors we really liked, and that doctor talked with me on the phone. She told me she would be the take-charge person for me, and that is exactly what she did, even though I later learned she was actually a liver specialist.

"After deciding against the clinical trial, we both agreed we would not go back to the same city again for treatment, that whatever was going to occur was going to occur at home for better or worse.

"Once we got back home the second time, the interaction with doctors and all the other medical and administrative people was wonderful. While we don't have all the specialists and equipment locally that we had in a big tertiary care center, what we do have is a group of warm, caring people who recognize you on sight and just get things done when they should be done. It took awhile for both Charles and me to appreciate what an excellent physician we have, as well as a caring person. Our hospital has 'You are a star' cards to fill out for people who are especially helpful, and over the years, I have filled out several. Just recently, I submitted one for the scheduler/jack-of-all-trades for our doctor. I had some trouble with the pharmacy at first as I was not in the mood to be denied anything Charles needed right then! The whole staff there now knows who he is and often asks how he is doing, and I have learned to manage his prescriptions better."

"Jim had a very positive attitude," states Judy. "That carried over to his relationships with his docs. The doc who had the most input into his overall treatment was his oncologist, and Jim received a lot of support and information from the oncologist's office."

Kathy describes her relationship with the doctors. "The one thing that was true throughout was that when I had specific questions about my husband's treatment, I did go to the doctors and got answers. Sometimes, these answers were not what I wanted in terms of certainty, but the doctors were always willing to address my concerns."

Michele agrees, "Our relationship with the doctors was good. I really liked his gastroenterology doctor and GI surgeon. They really seemed to care and laid out all the options well. My dad really liked all of them as well. His oncologist was good yet not specialized, so it was difficult to take his opinion as well as I would have liked. We did go see a specialist with whom he worked well. His radiation oncologist was great, very positive, and helpful. My dad and he spent many of the appointments talking about golf games and kids. As I mentioned before, two of the physician assistants were extremely helpful. I could call them and ask anything, and they would always have a great answer. Maybe they did not provide me with the one I wanted, but they were very intelligent and supportive emotionally and intellectually. My dad's surgeon actually came to a recent golf tournament fund-raiser we put on for my dad even though he never went into surgery and he was not a big golfer, if that says anything. This surgeon is just a great guy!"

According to Jennifer, "We had good relationships with all the doctors. I was the advocate for my mom, along with my stepfather. I didn't have contact with my father's doctors."

Melissa was very positive about the doctors at a leading health care facility. "They were absolutely amazing! They became like friends to us. They were invaluable. I will never forget the two oncologists who treated my dad."

FAMILY PICTURES

With Contributors and Patients (survivors and deceased)

Family #1

Family #2

From left to right: Ryan (contributor), Lisa (author), Husband, Fred (patient-deceased) and Sari

Front row, left to right: daughter, Teresa, granddaughter, Heather, author, Pat (contributor), Back row, left to right, son-in-law, Luay and husband Charles (patient-survivor)

Family #3

From left to right: Jennifer (contributor), mother Ethel
(patient-deceased) and Jim

Family #4

From left to right it is:
Ted (brother), sister Jackie (contributor), Joyce (patient-deceased),
Janet (sister), Vickie (daughter of Joyce), David (son of Joyce)

Family #5

From left to right: Katie, Megan, Brooke and Jen (patient-survivor)

Jen and Michael

Family #7

Family #8

Kathy (contributor) and husband Vinny (patient-survivor)

Judy (contributor) and husband Jim (patient-deceased)

Family #9

Michele(contributor) and father Greg (patient-deceased)

Family #10

From the left, there's Nick (brother), his wife Christine, mother Dolly (patient-deceased) holding Nick and Christine's daughter, Elliana, Deborah (contributor), and father Yeh-Yu

Family #12

From left to right: Lee, son; Lee's wife, Sandra; Jeanine (contributor), Cheri, daughter; husband Don (patient-deceased), holding Lee's dog, Ollie. The boy in front of Sandra is Cheri's son (grandson,) Devin. Sitting in the chair in front of Cheri is her husband, Heath. Sitting on Heath is Cheri's daughter (granddaughter,) Emma

Family #13

Left to right photo: Back row first (adults) Christopher (son), Janelle
(daughter in law) husband, Floyd (patient-deceased) Bette (contributor),
Craig (son), Deb (daughter in law). Floyd III (son), Carrie (daughter),
Gary (son in law), Shane (grandson), Brandon (between rows grandson
next to grandpa), Jeremy (grandson), Tiana (grandaughter), Tristan
(grandson), Camdyn (grandaughter), Beau (grandson)

Family #14

Melissa (contributor) and Dad,
James (patient-deceased)

Family #15

Sherri (contributor) and Mom,
Margie (patient-deceased)

Family #16

Len (patient-deceased) and Sue (contributor)

Family #18

Top Row: Grandson Greg, Oldest Son Jeff, Jeff's Wife Cherry, Daughter Laurie, Daughter-in-Law Jenny, Her Husband (our Son) Mark. Middle Row: Grand-daughters Kelley and Michelle, wife, Frayda (patient-deceased), and Aaron (contributor), Grand-daughter Paige, Lowest Row: Grandson Ali, Grand-daughters Sara and Holly.

Family #19

Cinzia's (contributor) Mom and Stanley (patient-deceased)

CHAPTER 4

Winning the Battle
but Losing the War

Restore a man to his health, his purse lies open to thee.

—Robert Burton

"I never liked roller-coasters," stated Lisa. "The disease was most like the ride I hated. As we experienced the ups and downs of pancreatic cancer, each of us began to live from scan to scan, test to test, doctor's visit to doctor's visit. It's difficult not to examine each report, hoping for some encouraging news. We call this the good news, bad news of the disease."

Good news/bad news

"The good news of the cancer is the overwhelming support, love, and prayers that we got from family and friends. The bad news is that we lived under a cloud of unknowing. After Jim had surgery and chemo, he had blood tests and scans every three months, and we worried about every bump in the road," explains Judy.

Jeanine agrees about support. "The good news is that Don's illness brought the immediate family closer to God spiritually and to each other emotionally. We were also able to see how much other family members and friends loved Don and wanted to support him in whatever way they could. Another good aspect of his illness is that it made us, his immediate family, more aware of the importance of good nutrition in maintaining good health." In his CaringBridge posting, he stated,

"Each night we go to bed with complete peace, and each new day brings hope. Whatever happens, we can say that some of these days have included the best moments of our lives, although they have ironically occurred during the worst time of our lives."

"The bad news of the disease is the pain it caused Don, both physically and emotionally. Because of his pain, his family and friends also suffered from seeing him suffer, and at times, he was not able to savor life as he had in the past. The worst news is that it took Don out of the life that he loved and enjoyed so dearly."

Deborah describes the ups and downs of her mother's experience with the disease. "Up—I e-mailed one of the leading pancreatic cancer doctors at a well-known hospital for pancreatic cancer and secured an appointment for my mother. Down—seeing how exhausted and physically/mentally drained she was from the three-hour trip to the hospital. Up—finding out about promising trials for which my mother was eligible. Down—hearing from the oncologist that he honestly thought my mother was too weak and should be treated locally.

I'll never forget one balmy summer day, when Gemzar was still causing my mother's CA 19-9 to drop. She had a wonderful appetite that night for dinner, and we were sitting around after the meal, just chatting. She looked so happy and at peace when she said earnestly, 'I think I'm getting better.' She spent the rest of the evening sitting outside with my father, enjoying the weather, and cleaning out one of the birdhouses, gently picking and prodding out old twigs with tweezers. Two days later, she was rushed back to the hospital due to a blood clot found in her leg."

From someone whose husband is currently fighting the disease, Kathy states, "The good news about the disease is that you don't die right away. The bad news about the disease is that you don't die right away. Look at the paradox. I mean, you want the person to live. You are grateful that he is living, but either over time or in periodic spurts, you begin to question yourself, 'What is the quality of life that he is living?'

Is going through all the treatments and what is connected to it worth it? When Vinny was first responding well to treatment and we saw his oncologist every three weeks, I remember her saying, 'We are going to keep doing what we are doing.' I remember thinking, 'For how long? For forever?' I mean, some people get to have treatment, suffer through it, and then it ends. There is a phase called life after cancer. For us, Vinny and I, the bad news is that we don't know what the future may bring, and that is true in a different way than someone who has, say, operable cancer. That is the bad news. The good news is that he is responding and our love has deepened. My days are defined by being happy simply because he feels okay.

"The ongoing challenges of not knowing about the prognosis are that it becomes harder and harder to make plans. You never know when (what days) the person will feel well and/or when chemo will begin again. Therefore, to plan a trip or an event even three weeks into the future feels impossible. Yet and this is an important yet, you learn how to try. You make a few plans and try to stay flexible and hopeful."

Pat, another survivor's spouse, writes, "Your mental condition after your spouse is diagnosed with an advanced adenocarcinoma of the pancreas must be just a bit like being bipolar. The mood swings have been so extreme that sometimes they were downright scary.

"Given the awful news that the best-case scenario with successful chemo would be six to nine months, I was as shocked and saddened at the thought of losing my husband as anyone would expect to be, probably not any more nor less than other people."

Pat continues, "We made plans to visit our daughter and family for two weeks at Christmas, take a cruise to the Amazon at the end of January, and go with our daughter and her family to Hawaii in March. Ten months ago, we were quite certain he wouldn't be alive to do any of those things, so we were feeling very good about doing those things. Most of the time, I felt pretty normal. And we were both pretty pleased about the trips. But there were the crazy days. I cried off and on all day at every little thing and couldn't begin to tell anyone why. I knew what triggered each crying jag, but why that day and why not some other day is a complete mystery.

"Then our world turned upside down again in a much better way. A CT scan revealed that the tumor was no longer visible. There was 'fuzziness', as the oncologist described it, around the head of the pancreas, and he thought there was probably a malignancy remaining since cancer markers were still a tiny bit elevated. To our utter amazement, he now thought radiation might be an option. When we saw the radiologist, he said the CT scan might be showing malignancy or might be showing scar tissue, but he thought Charles might be a good candidate for radiation therapy in addition to the chemo. He outlined risks first, including some very serious risks. We weren't real concerned when he talked about life-threatening risks. After all, pancreatic cancer is going to be fatal if treatment doesn't get it all. After that, he said there was a 15 percent chance of a cure. Charles and I looked at each other. I think we both couldn't take in the word *cure*. No one had ever suggested that might be remotely possible. Charles asked him to define cure, and he said, 'You live five years or more and die from some other cause.' The doctor then had to leave the room for a few minutes to do something with the equipment for a patient being treated. When I turned toward Charles, he had a look on his face I've only seen a handful of times—when we bought a house we really wanted, when he first saw our baby granddaughter."

Then there are the true realities about the disease. "There was no good news in the disease for our family. The bad news was that our sister, wife, mother, and friend was going to live with the disease and battle it all the way," states Jackie. "The challenge was the unknown. It was so hard to know what the right decisions were and what to do about everything."

Bette agrees, "The bad news was there was no treatment for this disease. It was a death penalty upon being diagnosed in most cases. It was difficult to accept the fact that he was not going to be around long."

"The situation can also change in an instant," Michelle states. "I really felt negative when we got the news that they did find that the spots on the liver were growing. Therefore, they were not going to operate. This basically sealed his fate, and it was horrible. After that, it was really difficult to fight and remain positive."

"Not knowing what is happening is the most challenging aspect of the disease," Lisa states. "By summer of 2008, only eight months after the diagnosis, Fred's tumor decreased by 80 percent due to radiation and chemo.

We lived from scan to scan, analyzing every report and asking detailed questions. After six cycles of Gemcitabine, the CT scans showed no living tumor cells. However, we didn't know what to do next, and second-line treatment was not available in the hospital where we went. Therefore, off we went to hospitals with national reputations. We spent our summer gathering facts and getting treatment options. They all seemed to point in the same direction—clinical trial. And Fred wanted no part of it. So we were at the top of the roller coaster, about to zoom down."

Ryan agrees, "After the tumor had shrunk, Dad was considered cancer-free, as the doctors called it. After we found that out, we were ready to take the next step. After consulting the doctor, we were told that there wasn't a next step to take, and we had to pretty much take my father in for CT scans every three to six months. The reason for the CT scans was to see if the cancer had come back and if the tumor had started to grow again. The doctor's role seemed to be that they had done all that they could do, and there wasn't going to be another step. This was not an answer that was acceptable."

Sometimes, you just aren't connected to what is going on. Jennifer explains, "With regard to my father, I did not know much about ups and downs as he was in Colorado and I was in Virginia. I would go visit, but it was hard to tell how he was doing on the phone. In the end, I didn't know he was dying. His wife kept it from me, and I didn't get a chance to see him one last time to say good-bye."

The doctor's role in staying positive

According to Bev, "The positive aspects were our faith in my husband's oncologist and knowing about the successes that other pancreatic cancer patients—who were receiving the same aggressive chemotherapy—were having. Because my husband was treated by an oncologist who specialized in pancreatic cancer, at each appointment, we were able to talk to other patients and their spouses and discuss the issues with which we were all dealing.

"My husband's oncologist was terrific, always emphasizing the positive, such as the early and significant decline in my husband's CA 19-9 blood levels after he started chemo. My husband was in good physical shape when he was diagnosed with pancreatic cancer. The term that doctors use to categorize the patient's level of physical stamina is performance status, and at the outset, my husband's was excellent. The doctor urged my husband to continue to keep up his strength with walking, either outside or on the treadmill at the gym, whenever he felt up to it."

Sherri described the treatment of the medical staff to her mother. "While she was in the oncologist's office getting chemo, the doctors and staff couldn't have been any nicer to her, offering to do anything she needed if it might make her any more comfortable (brought a drink, pack of crackers, a warm blanket, turned off the overhead lighting, and pulled the door shut so it would be quiet, etc.)."

"The doctors were always positive and kept encouraging my sister to keep fighting. I personally felt it was too much because she began living to please the doctor. At times, the doctor would encourage her to enjoy her life and do the things that she wanted to do but also told her that the chemo is the reason she is doing so well. She had faith in the doctors and believed that the doctor could cure this disease. Consequently, she spent two-and-a-half years going through chemo and only getting about two good days each week. I think it was to her detriment," Jackie states.

There were also mixed messages from medical professionals. According to Jeanine, "As I mentioned previously, the oncologist did not give us a prognosis but also was not positive in his dealings with us. He had been in private practice for over thirty years, retired, and was contracting his services to the VA on a part-time basis. His attitude was matter-of-fact and not particularly hopeful, even though he never directly said Don's disease was terminal. In comparison, the radiation oncologist was younger and much more hopeful. So hopeful, in fact, that the palliative care doctor, who was also Don's primary care physician, became irritated that he was giving us unrealistic hope. However, we were thankful for the hope that doctor gave us. Our spirits were always lifted after an appointment with him.

"With the exception of the radiation oncologist, all other doctors were sending us the message that we should not hope for Don to live much longer, and we struggled to fight against that negative influence. In addition, they discouraged us from considering alternative approaches and discounted all the hopeful reports we would bring in about the use of the alternative approaches. Even when Don had recovered to feeling almost normal in the summer of 2007, they would not acknowledge that the alternative treatments he was using had anything to do with his recovery. Thankfully, they did not discourage us in a heavy-handed manner, but they did get their message across to us."

More conflicting messages are explained by Deborah. "My mother's doctor would throw in these halfhearted statements beginning with 'When you get better . . .', but I almost wish he hadn't done that. He seemed pretty cold and standoffish, but I only visited him one time. The one conversation that made my mother partial to him was when my parents told him that I planned on putting my career on hold and moving home. I don't know exactly what they talked about, but the end result was that he gently convinced them that it wasn't the right move. That he had been so open and willing to share his experiences and opinions was really valued by my parents."

A close relationship with the oncologist is described by Pat. "The oncologist is supportive and honest with us. We only see him every four weeks, which, in retrospect, was not often enough at the earlier stages of treatment. There were just too many unanswered questions between visits. He treats us as intelligent human beings and does a good job of balancing encouragement with realism. In the early stages, he was much better able to articulate some things my husband needed to hear than I was, and that helped me a lot. My husband had some rather strange ideas at various times about what some symptoms might mean. When I could not get through to him or did not feel I should even try, I would ask the question of or mention it to the oncologist during a visit, and he would explain the answer in ways my husband was more likely to accept. That helped me emotionally more than I can describe."

Bev agrees about the support of the oncologist. "The oncologist answered all of our questions and was available by phone, either in the office or after hours, for the issues, usually related to how to handle chemo side effects that inevitably occurred."

Then there are others who did not have such a positive experience with the role of the oncologist. "I always felt that the doctors wanted to do their best, but I wasn't sure whether they were the experts we really needed," Lisa says. "I felt they didn't have answers to my questions. Maybe there weren't any, but I wanted and needed to hear more."

Sue agrees, "We didn't care for the oncologist in charge of our case, but there were several other doctors who were compassionate and with whom we related more positively."

Finding support: relationships with family and professionals

"Leaning on friends and anyone who would listen was my lifeline," explained Lisa. "I went to my caregiver's group religiously. However, Fred's experience with a social worker was not positive, and he never went back. He continued to put on a brave face and was not a complainer. But I knew how angry he was that his life was being taken away by this horrible disease."

Ryan states, "My relationship with my mother grew exponentially because of how honest she was with me throughout the entire process. I never once sought out any professional help with this matter."

Bev explains the importance of family. "Our family is small and very spread out across the United States. All of the family members were very supportive when they were told about my husband's cancer. Because there is such an extensive history of cancer in my husband's family—including an older brother who died of pancreatic cancer in 1989, just before his fiftieth birthday—the news of my husband's pancreatic cancer, while very upsetting to other members of the family, was not as much of a shock as it would have been in a family in which cancer was very rare."

"I had family and friends around to keep me going," states Bette. "I tried to make each and every day a good one, so I had no regrets as to our last days. We went on a cruise, traveled out East for two weeks in our new motor home, and took a trip around Lake Michigan. I needed the support of family and friends and also of my husband telling me I would be okay, and this is not what he had hoped for, but these were the cards he was dealt. He was not afraid of dying.

"Our family relationships grew, and we were very close and supportive to each other. It was a time where everyone was positive, and we all needed each other to keep on going. We all were positive and happy."

Jackie describes the support of her family. "I have a good partner in my personal life that took care of everything while I was away taking care of my sister. He was my angel, along with my other two siblings. I would never have gotten through this time without their support.

"We grew up in a dysfunctional family, and the four of us children were always extremely close. We would always be there for each other, and this was no exception. We did not use professional help. We had each other. We are very loving and caring for one another."

"Finding emotional support was challenging," according to Pat. "The emotional support I needed simply did not exist. I only have casual friends where we live. All of them were supportive, and many offered help. Neighbors invited me out to lunch to get me out of the house, and it was a help. But I didn't have anyone I could pour my heart out to except my daughter. She was in the process of losing her father. It wasn't fair to burden her with supporting me too. She had been through recent brain surgery herself and lost her college roommate and best friend to breast cancer only two years before. This time was certainly not easy for her either. Sometimes, I would e-mail her in frustration, but generally within what, I was confident she could handle. One time, though, I just came unglued and called her. How she managed, I'll never know, but she managed to talk with me for about an hour without coming unglued herself. It didn't solve my problem but kept me sane for another few days.

"Practical support was never a problem. People offered to do things, most of which we really didn't need, but there was always someone I could call on if there were something practical I needed.

"My next-door neighbor—whose husband has another ugly terminal cancer, mesothelioma—and I persuaded our local cancer resource center to start a caregivers' support group. It is helpful, but we haven't been able to get the word out enough to have a large enough group to be sustainable, so I don't know how much longer we will be able to keep it going. If only Sue (the neighbor) and I show up with the facilitator. It isn't as useful as it should be. We live next door to each other. We really don't need a formal group to talk with each other on this subject. There are some odd moments. One time, the facilitator asked each of us what we were doing to sustain ourselves. Sue had recently managed to get away with women friends for a couple of weekends while her sons stayed with her husband. The only thing I could come up with was that I had had my long-delayed back surgery (which was really important as I can now stand around and I can take walks). But when gently pushed, there was nothing else I could come up with. I told the moderator, 'You just do what you have to do.' I became very emotionally fragile again for a week or so."

Kathy describes different types of relationships among friends. "The things needed were a variety of different kinds of friends—supportive friends, nonjudgmental friends, those who could hold you and not try to fix it, which was impossible, those who helped you do one or two normal things, those who brought food, and those who simply offered a hug.

"Relationships with family members are/were up and down during this period and continue to be so. It depended upon my level of fatigue and my sense of their level of insight and sensitivity. Slowly, I learned (and am learning) whom to trust, whom to ask for help, and whom not to ask for anything.

"I did seek professional help during this time. In addition, the family was offered family therapy provided by the hospital where my husband is being treated. It helped at times, was a sounding board, and a place to prepare and vent. At times, the family therapy made me feel worse, and at times, it helped a little. None of it takes away the pain, simply offers a way to try and communicate about it."

Deborah also describes the important role of friends. "I kept up my reserves by unloading a lot of my emotional stress onto my good friends. They were my saviors. Several of them had lost good friends or family members to cancer. They knew how I felt and made themselves available to me whenever I needed them. I also started a journal to help express my feelings. In addition, I spent a lot of time trying to spread awareness and raise funds for Pancreatic Cancer Action Network with the River to Sea Relay race.

"One of the few positives from this experience was how much closer my brother and I have become. He is seven years older than I am, so after he left for college, we slowly drifted apart. He and his family live about three hours from my parents, so I saw him much more often after our mother got sick. Even when we didn't see each other, we would e-mail, call, and text message much more often than before. I've always been close to my father, but seeing how he behaved in this situation was really awe-inspiring. He and my mother have never had a perfect marriage, but my mother could not have asked for a better caregiver."

Judy describes the various kinds of support that helped. "We had a lot of support and prayers from family and friends. That's what we needed and what we got that mattered throughout the whole experience.

"We had been seeing a couple's counselor, and we continued seeing the counselor throughout most of Jim's illness and recovery. It helped us to deal with the stress of the unknowing. We also joined a couple of pancreatic cancer support groups and attended the local Pancreatic Cancer Action Network symposiums when they were held in our area.

"Our relationships with family, both immediate and extended family, were very strong and continue to be very strong, even after Jim's passing.

"As I stated earlier, we had been seeing a couple's counselor, and we continued to do so throughout most of Jim's illness and recovery. We also joined a couple of pancreatic cancer support groups and attended the local Pancreatic Cancer Action Network symposiums when they were held in our area."

CHAPTER 5

The End and After

The fear of death follows from the fear of life. A man who lives fully is prepared to die at any time.

—Mark Twain

Sounds fine in theory. But in actuality, not many of us are prepared to lose those we love. Some of us began the planning process when we realized the end was near. Others continued to hope for miracles and believed that things would get better soon. The bottom line is that this was the most difficult time for all of us, watching our loved ones decline, trying to stay on top of pain, and wondering what would happen next.

Planning for the worst

Lisa explains the decline in Fred's health and the planning that she initiated. "By the time the New Year had arrived, Fred was very sick. He was yellow, bloated, and in even more pain than ever before. By the middle of January, we contacted a doctor in New York City and had a consult. After some CT scans, we were told that he needed to go immediately to the hospital. While in the emergency room, his liver numbers were off the charts, and we were told he was in liver failure. He was admitted, but we had time to talk while waiting for a room. We agreed that if the hospital could do nothing to help him get better, then we would go home. He stated that he did not want to die in the hospital. He stayed in the hospital for six days. After many tests, blood clots were discovered in his heart, lungs, and liver. Attempts were made to remove the clots with medication, but we were told that there was very little hope of eliminating them at this stage. I called our children and Fred's sisters, who all came to visit.

"Fred and I had also discussed how we both felt about funerals and cemeteries. We both agreed that we wanted to be cremated and have our ashes spread in a special place that would mean something to us.

"While Fred's testing was going on, I stayed at the hospital and began making phone calls. I found a small funeral home that advertised personal memorial services. I spoke with a funeral director and told him our situation. He was very kind, and I felt better knowing that I had a plan if I needed one."

From a son's point of view, Ryan explains his perspective at this time. "After the cancer had spread throughout my father's body, we decided to take him home and spent the last couple of days with him together. Right before we left the hospital, we discussed what would be the best choice of action, and we were all in agreement that there was nothing more that could have been done. My feelings at this time were pretty straightforward—I was upset with all the doctors that took care of my father because all of them gave up on him after he had gotten the chemo. They all said the exact same thing—that there was nothing that could be done."

Sherri agrees with the planning process. "Mom and I had talked about certain things up front, so we both would know what she wanted and so that I could carry out her wishes to the fullest extent possible. She wanted to avoid going into a nursing home if at all possible, which we were able to avoid. While she was still able to do it, she pulled out her nicest sleepwear, ironed it, and hung it out so that it would be ready should she need to be dressed for company. She already had her cemetery plot purchased, but no arrangements had been made with regard to the funeral home. I would not have allowed her to go alone for anything in the world, but it was with a heavy heart that I went with her to the funeral home and made the preplanning arrangements. She did not want to be involved in all of details of the actual service and said she would leave that up to me, but she made arrangements for what mortuary services she wanted from the funeral home, including picking out her casket. The humor we got from that was that she asked me if I thought it was too flashy. Needless to say, I told her no."

Deborah explains her family's situation. "Most of the big decisions were in the hands of my mother and father. They jointly agreed to hire a caretaker instead of asking me to move home, to be treated locally instead of at a teaching hospital, and to join the hospice program. My decisions were more trivial, but some were just as difficult. How often should I go to visit my mother? Should I acknowledge my mother's increasing weakness, or ignore them like I think she'd like me to? What should I cook tonight?

"My mother and father were very open to my brother and me about the decisions they had made. While we made it clear to them that we didn't always agree, we respected them. Looking back, maybe the decisions I wanted them to make were for selfish reasons. After her initial diagnosis, my mother told my brother that she had lived a good life, was blessed, and was happy with what she had. Maybe that's why she was able to go as peacefully as she did. And yet we still tried to push her to pursue more aggressive treatment, to 'not give up' so we could continue to have her there."

Hoping for the best

Michele explains her father's attitude and feelings toward death. "My dad and I didn't really get a chance to talk about many of the decisions because we weren't expecting his death so quickly. It was really hard for my dad to open up about what was going on and deal with the reality that inevitably he was going to die. He did say he was scared to die in two different conversations we had, but we didn't talk or make decisions about things because his death came on so fast and unexpected. We did talk a month before he passed about one thing. I asked him with a soft preface that I wasn't saying he was dying yet, but I asked if he wanted to be buried next to my mother. My mom had passed away several years ago, and it was a tough ending, with my dad and her not on the greatest of terms at the time, but he loved her so dearly. He said, 'I have been thinking about that a lot lately.' We chatted a minute longer and then he quickly changed the subject. I didn't get a real answer from him at the time. I found out when he passed that he had bought a plot with her, so that was the answer. Yes, of course he wanted to be with her. He loved her so much."

Judy speaks about hoping for a miracle to happen. "We always had hope that Jim would somehow beat the cancer. After his diagnosis of recurrence, his oncologist put him on Gemzar and Tarceva, and he did well for many months. After the oncologist thought that the combo was not working, he sent us for clinical trials at the University of Chicago. Jim was scheduled to start a clinical trial two weeks after he died. When he signed the papers for the trial, we told the doctor that he was doing it for the chance to help other pancreatic cancer patients and perhaps benefit from the trial himself.

"We were *not* thinking of end-of-life issues until I had to take Jim to the hospital for the last time. At first, we thought that he was just dehydrated, but they wanted him to go for an MRI. His kidneys were not working well, and the kidney doctor came in to speak to us. Jim was up and talking and lucid. He ordered lunch and ate it, and people came to visit, and he talked to anyone who called on the phone. His primary care doctor kept coming in to update Jim, his mother, and me on the results of his blood work that they were constantly checking. His doctor came in late in the afternoon to tell us that what they were doing wasn't working, that Jim had a serious blood infection, and while they were still treating him with antibiotics, they needed to know his wishes if he went into cardiac arrest. I spoke first. 'No heroic measures, no breathing tubes.' I don't know why I spoke first, but Jim affirmed it, and we both agreed 'no heroic measures.'

Jeanine explains the faith she had that Don would rally. "I'm not sure I could say that we actually made decisions. Since we both were in constant prayer for divine guidance, I feel that decisions were made for us by our Creator rather than our making the decisions. But others may see it as decisions that we made. An example is the decision to put Don on hospice in February 2008. Don's doctor was located at a government hospital in Portland, Oregon, which is sixty miles from our house. She had been urging us to put Don on hospice ever since he was diagnosed in 2006, but we refused because we were determined to believe that Don would be healed, and we saw him going on hospice as an acceptance that he was dying. The doctor tried to get us to see it as a way to get local help when needed, rather than having to always drive to the VA Center. After two years of frequently making that trip, we both were weary of it. Don had grown to hate the Center and everything connected to it, so he especially resisted the trip. In February 2008, Don had yet another gallbladder attack that produced excruciating pain. When I called the hospital, I was told that we would have to make another trip in to get help, or we could put Don on hospice and a local hospice nurse could come to our aid whenever needed. At that point, we were ready to have Don go on hospice but with the expectation that he would improve (as he had in 2007) and eventually get off hospice.

"As Don's condition continued to worsen rather than improve, we still kept the mind-set that he would eventually rally and never talked about what was happening in any other way. Even when he developed ascites, lost his appetite, and gradually weakened, we still approached everything with the attitude that he would eventually rally. I remember talking to the hospice nurse about cognitive dissonance, meaning that my mind was struggling with holding on to the belief that Don would rally, while at the same time seeing all the signs that he was declining. As the pain medication increasingly affected Don's physical and mental abilities, the less we talked, and the decisions as to what to do seemed to make themselves (i.e., be made by our Creator) rather than be made by us."

Sue had a feeling that Len wasn't going to get well. "On Monday, May 8, 2007, Len went into a care center. It was a suggestion from the hospice nurse, who felt I should take advantage of the five days offered by Medicare to give the caregiver a break. I decided to go ahead and chose a care center that was about a mile away. It was a small one, and I had heard good things about it from others.

"As he was wheeled out in the wheelchair by my friend, I had a strong feeling that he would not be coming back to our home.

"One of the things I was planning on was getting more sleep since it had been hard to sleep with his breathing so loud and being concerned for him. I ended up going over to the care center quite often and spending a lot of time there, even tried to sleep there on a chair or the nearby couch, but his breathing was so loud it was hard to sleep."

Carrie describes her father's feelings about death. "I have to say that my dad did not like to talk about the end of his life. I suppose most people wouldn't like to do that. But I have heard of some who wrote their own obituary, wrote something to be read at their funeral, or planned every little detail of how they wanted to be remembered. I guess, in secret, my mom and I would talk about if dad died, where we would have a funeral, but we never had it in us to talk about the details. See, we were always in denial and had hope until the very end. We never thought we were going to lose

Dad. Never. It was so traumatic to even comprehend it. We just never even talked about making plans. We knew the statistics and the survival rate. We knew what the doctor was telling us, and we knew Dad was changing, but we never thought about a casket, a funeral, a cemetery, or the final outcome. At least I didn't. (I guess Mom and Dad did go visit the cemetery where he had his plot that was next to his father.) I do believe I didn't think about this as much as other people going through pancreatic cancer because my dad never really showed many symptoms during his whole illness. Just when he began to start to feel ill, he went downhill rapidly, like in two weeks. He never lost weight, never took pain medication, and never sat in bed resting because he couldn't function. He was riding his snowmobile twelve days before he died. He was going to NASCAR races, driving his own motor home from Minnesota to Texas two months before he died. He was hearing that he was dying, but he kept living until he died. The week he died, he was heading from Minnesota to Oklahoma to a big car-race event he loved to watch. He was packing on Sunday, and Monday morning, something happened. He couldn't breathe, and he couldn't catch his breath. He and my mom decided he'd better go into the hospital, and that was his last week of life. His kidneys failed."

Melissa states how quickly things can happen. "We really didn't expect my dad to pass when he did. We took him to hospice only to get his pain under control. I can still remember driving him to hospice. He turned to me and said, 'You aren't going to leave me there forever, are you?' I turned to him and said, 'No, Dad, I wouldn't do that to you.' I sometimes regret I ever said those words to him.

"Dad was in hospice about three days, and the doctor came in and talked with us all. He told my dad he didn't think there was much more they could do for him. He was sorry, and he thinks he should talk to everyone with whom he wants to talk. My dad responded and said he hoped the doctor was wrong. The doctor then asked if he wanted to stay there or go home in his last days. My dad said he wanted what we (my mom, brother, myself) wanted. But we all said it was up to him. He decided it was best to stay there. He didn't want to burden anyone."

Aaron discusses the added time that his wife had due to her experimental treatment. Frayda survived for a period of fifteen months after her diagnosis of metastasized pancreatic cancer, even though the original prognosis was three months. We feel that the one-year extension can be attributed to the effectivity of the experimental monoclonal antibody.

"The quality of her life during the first twelve months, while only fair, was good enough to permit some travel during chemo breaks and to obtain fairly frequent visits from family and friends. It was only in the last three months that her condition deteriorated rapidly. She developed pulmonary hypertension, apparently from the Gemzar that was administered along with the monoclonal antibody as part of the phase II study. She had to be removed from the study, and what followed was a series of failed attempts to find another effective treatment.

"When her last CT scan showed extensive spread of the cancer, it was decided by me, in conjunction with her oncologist, to stop the curative treatments, which were completely ineffective, and start palliative care. She was transferred to the hospice wing, where she was kept comfortable but lasted only about three days.

"In discussing the change in care with my children, it was very clear that there was no other choice than the one we took. They were very upset but, at the same time, quite supportive.

"When it was clear that hope was fading, I made arrangements for the funeral, without discussing it with Frayda."

Families unite

Lisa states, "I spoke with both kids about what Fred wanted. They felt that it was his decision and would respect whatever he wanted. Though his sisters were Catholic and may have had different feelings, they respected his wishes."

Melissa agrees, "My family is small (Mom, brother, myself). My dad has brothers and sisters, but they did not keep in touch. However, when they told my dad he was in his last days, we called all family members. They all came to say good-bye. Everyone was on the same page. As my dad knew he was dying, most of the decisions were left up to him, and we supported him."

Sherri describes her relationship with her mom. "Everyone was on the same page in that I told Mom I would do whatever she wanted to be done. Her decisions were logical, well thought-out, and made well in advance when her frame of mind was still sound. Short of something just becoming physically impossible (which nothing like that happened), her wishes were going to be carried out."

"My family agreed it would be a good idea for him to go to the care center. My granddaughter, age five, was very close to Len and came over there and helped feed him and keep him company," states Sue.

Carrie explains about letting go toward the end. "I have three older brothers. They were given the chance to come to doctors' visits, invited to stay at hotels during treatments two hours away. Sometimes, they did but not as much as I did. I am the youngest and only girl. I was Daddy's girl. I do remember having the feeling that I never had to get more of Dad before his death because I had always been close to him, and there was never anything left unsaid or unfinished. My parents were a part of my daily life, and we were always up on what was going on in each others' lives. So I was shocked with myself that when the end was drawing nearer, I actually let go a little bit. I stepped back and let Dad have his space to do what he needed to do and when. It was so easy to need him to make my mom and me feel better because he was dying, that I could see he was getting tired of doing that. He gladly did comfort us and console us, but get real. Who wants to have to play counselor to your wife and daughter day after day because they are so fearful of living life without you? He needed his time and space to go where he wanted and do things that he might never have done. I had my time with him *all* those years. Now it was time for him to go to dinner with old high-school friends and explore what he wanted to explore. I could have been so needy, but thankfully, I had a friend who helped me see what was important and just talked to me whenever I needed to talk about it, which was much of the time."

Families keeping quiet about the end

Michele explains her family's reactions. "We didn't talk about it with anyone. My dad and I are mainly each other's family. He had many friends and was close to his brother, but we didn't bring up the topic with anyone.

I don't know if we would have, but again, we weren't expecting the quick end."

Judy describes the circumstances surrounding Jim's last days. "Jim's mom was at the hospital with us all on that last day, and the rest of his family came to the hospital. Jim talked to them as they came and as he was able. His sister was the last to come. She had to drive from a distance and came when we were ready to have a short communal prayer with Jim in his room.

"No one questioned how ill he was and the fact that he was probably going to die."

"Our two children also talked in terms of Don being healed, not dying, so there was no conflict in the decisions that were made," explains Jeanine. "They were supportive of his going on hospice for the reasons previously cited and not because they had accepted that he would die soon. In many ways, it was like the proverbial elephant in the room. Everyone knew it (the possibility that Don might die) was there, but none of us acknowledged it. When friends came to visit, I could see that they thought Don was dying, but I always felt certain that they were wrong. At the same time, a part of me did acknowledge that it was possible that he might die since everyone will die some day."

Coping with the end

Lisa explains her feelings toward the end of Fred's life. "I went from logical to very, very sad. It is so hard to finally give up the fight that you have been doing for so long. I also knew that Fred really wanted to live. A dear friend came to visit, and we listened to stories, trying to keep from thinking about Fred's situation. I slept in his room every night. I made calls during the day. I tried to take control over whatever I could because I couldn't change the rest.

"When we finally went home in an ambulance, we talked about that the time had come to call hospice. They came within a couple of hours, and we went through all the detailed instructions. The whole family started to make sure Fred was comfortable, not in pain, and had whatever he wanted and needed. That included all of us running around to buy comfortable sweats, a shower chair, mats for the bathroom, pillow top for the bed, and heating blankets to keep him warm.

"One important thing that I did when Fred was coherent and awake enough was to have him remember and talk about his life. I had bought two books for our children that contained questions about Fred's childhood and memories. All week long, I wrote answers to the questions to which Fred responded.

"The most important priority was that Fred would not be in pain. We were all there when Fred took his last breath. He was peaceful. Part of me was relieved that he was finally at peace, and the rest of me couldn't believe that he was really gone. We contacted hospice and waited for the funeral home as instructed."

Ryan explains about the decisions that were made. "Decisions for the funeral were made mainly by my mother. We picked out the place that it was going to be held, and my mom tried to include my sister and myself, but none of it really mattered to me. All I wanted was for it to be as seamless as possible and have it be everything my mom wanted no matter what. After the funeral, I felt that my father deserved to live so much longer than he did for everything he had done."

Sherri states that she did what her mother wanted. "As most of these issues were discussed and planned out enough in advance of Mom's passing, I was simply going through the motions at this time, listening to Mom, crying with her, talking with her, and assuring her I would be there for anything she ever needed. When we approached the dreadful time to implement these decisions, I think I was still going through the motions, saying, 'Okay, Mom said to do (so and so),' so we did it. Bless her heart, my sweet mother fought this horrible monster for two years and four months before succumbing and losing her battle, and as hard as it is for me to have thought this would have happened, I was still very much in shock when Mom passed. I expressed very little outward emotion. It was the same as any other day before it. We were just doing things a little differently than the things we had grown accustomed to, very mechanical."

Michele states that she was falling apart. "I was devastated, but in the role of staying strong for my dad, I was trying to make everything as great as I could. I would cook, sort medicine, and stay up with him or near him at night. I just wanted to make sure he was okay. I was so scared too. A month before my dad died, he fainted from blood loss, and I had to call an ambulance. After a short stay in the hospital, we continued life, and I was so scared it would happen again. I was on high alert. I felt overwhelmed and helpless. I would e-mail and call doctors/nurses, asking for options if it happened again. I was a mess."

Judy explains, "I was devastated but relieved that he didn't have to suffer the indignity of the side effects of pancreatic cancer. He was so sad that he couldn't eat the things that he enjoyed to eat. He had too many digestive issues with his recurrence."

Jeanine describes the feelings she had at the end. "In retrospect, I think I was numb most of the time. I focused on searching out and doing things that I felt would be of benefit to Don, and that kept me from feeling the agony and panic I know I would have felt if I let myself think of Don dying."

Carrie remembers the last days of her dad's life. "Dad died a horrible death with his kidney failure. He was in great pain and very uncomfortable. The hospital was not very sympathetic to his needs and comfort. I was very disappointed and sad to have had to experience this. My dad did not deserve to die as he did, in such pain, in this day and age of medicine. Due to lack of motivation and a crowded hospital, the staff was careless and selfish. Lack of communication and laziness does not have a place in a hospital that should be making their patients' last hours and minutes as calm and peaceful as possible."

Melissa describes her ambivalence at her dad's last days. "I had so many mixed emotions. I felt so sad to see my dad going through the pain, but I still didn't want him to die. So I felt selfish at times that I wanted a man who was in so much pain to endure the pain to be here for me.

"I can tell you in my dad's last moments he knew without a doubt we were there. My belief was also he decided when to take his last breath. I will never forget this special moment (kind of sounds strange). I believe my dad gave me a beautiful gift in his last seconds alive. It happened like this: he was having problems breathing. It was apparent. My mom said to him, 'Dearie (because that's what they called each other), if you have to go . . . we will be okay.' I broke down in tears because I believed that my dad needed permission from her to leave . . . I knew that's what he needed to hear. I then followed her comment and said, 'Dad, I love you!' He squeezed his eyes together really hard and took his last breath . . . almost to say . . . good-bye. It was with my mom, brother, and me all by his side."

Celebrating our loved one's lives

Lisa explains the details that were important to her. "At the funeral home, I asked for everything that Fred and I had spoken about for years. I wanted a memorial service that would celebrate Fred's life, not death. I also contacted the nondenominational minister that had conducted our twenty-fifth wedding anniversary. During the next couple of days, people came to visit—nurse, social worker, minister, and some friends. Fred was in and out most of the time. He was very tired.

"At the funeral home, we were also asked about memorial cards, and it occurred to me that it would be wonderful to have a photograph that Fred had taken, since he was a gifted photographer. We were also asked about the ashes. Our daughter and I agreed that it would be nice to have some ashes put into a necklace. She ordered a heart. I ordered a dove (symbolic of the dove that always was on top of our solstice tree, and I ordered two crosses for Fred's sisters).

"I asked about music, pictures, etc. The memorial service I had chosen encouraged us to make the service personal, and they were willing to do whatever we wanted. I felt relieved that it would represent Fred's wonderful life.

"The most important priority was that it was a celebration. I took all of Fred's plaques, certificates, pictures and brought them to the funeral home. I had already made a video of Fred's life, including Fred's childhood pictures that I asked his sisters to bring. I also worked very closely with the minister who would be leading the service. I wrote my eulogy and asked if the children wanted to speak. I began contacting everyone I could think of to let them know that Fred had died. My office had sent out an e-mail and had posted it on several LISTSERVs, so many people knew. I wanted to feel that people who had known him and worked with him would share in the celebration of his life."

Sherri describes her mother's role in planning her own funeral. "Mom had paid for the funeral and already had her cemetery plot, so we were left with the minor details of the service: picking out her clothes, jewelry, fingernail polish, and lipstick, etc. Mom always, always, always wore nail polish and lipstick, so there would be no reason why she wouldn't now. One of my brothers and I went to the funeral home to make the actual arrangements. My other brother had just driven down to Alabama the day Mom passed, so he was driving back while we were making the arrangements. I asked him if he had any preferences to anything, and he told me that whatever we planned would be more than fine. We hadn't given these minor details much thought until the morning of the day we were going to the funeral home. There was no point in thinking about these things until we had to do so. We just sort of made arrangements typical with what you would see in the Baptist faith. By that, I mean things like 'Amazing Grace' and 'How Great Thou Art' would have to be sung, etc. We kept it fairly simple, but I was told by others it was very classy and that they thought Mom would be pleased with how it turned out.

"The main priorities were to make sure that Mom was portrayed as the elegant, classy Southern belle (as she was described) that she was. Her clothing matched, her jewelry matched her outfit, and her makeup was complementary. Everything had to be just right. When picking out a memorial verse to print in her bulletin, you had to read it and automatically think of Mom before it was picked. Her appearances and wishes held the utmost importance to me."

Michele explains what was important to her. "The morning my dad passed, I called my uncle who is a funeral director in Fresno. He gave me the phone number to call the funeral home in town that he trusted to start the process. He came that day, helped me set up the appointment with the funeral home, and we went to the appointment together. Many of the decisions we made I either didn't care about or I surprisingly had strong feelings about them right away. It was bizarre. There was only one casket that worked. I didn't want any carnations at the funeral. I wanted a private burial and didn't want people to watch him go into the ground. I only wanted a few friends and family to see him, no open casket. I wanted songs that represented my dad, not a lot of Christian stuff. We are Catholic but didn't go to Mass that much. I wanted funny stories that truly represented my dad, and I wanted to speak to proclaim my love for my dad and how he was the best father in the world. I wanted to make him proud because I was the center of his universe and everyone who knew him knew that. (He *always* talked about me!) I wrote my speech the night before, and I cried the whole time while writing it, but I gave one of my best speeches at the funeral without fail. I know he was proud."

Judy emphasizes the priorities she kept in mind while planning the arrangements. "I made the major decisions with input from the family.

"Jim wanted to be cremated, and I wanted to accommodate Jim's coworkers so that they could pay their respects at a wake. He had worked at the same company for over thirty years. Since Jim died on an early Saturday morning, I felt that the wake could *not* be on Monday because his coworkers would only learn of his death on Monday. So we had the wake on Tuesday and the funeral Mass on Wednesday. Many of his coworkers came for the wake and funeral."

Jeanine kept in mind everything that Don wanted. "Before my mother died in 1976, she was adamant about not having a funeral. She said, 'If people care about me, they can come visit me before I die.' Don and I agreed with that philosophy and told each other that we didn't want a funeral or memorial service if one of us should die before the other. Don reemphasized that sentiment after he attended his father's funeral in 1996 and felt he had been turned inside out emotionally. He didn't want his family to go through that kind of misery. After his diagnosis, he called the local funeral home and made arrangements for his own cremation without much discussion with me. All I remember him saying is that he wanted that done 'just in case it was needed at some point in the future.' I can't explain why we didn't discuss doing the same for me, but we didn't. Again, the elephant was in the room but was ignored.

"When I was young, my parents bought cemetery plots for themselves, plus two for me and two for my sister. Even though Don and I were living in Oregon when he died and the plots are in Washington, we always assumed that we'd be buried in those plots. Both my parents and his parents are buried in the same cemetery, plus my brother-in-law. So after Don died, I made arrangements to have his ashes buried there and told relatives in the area that they were welcome to join my children and me when we buried his ashes. The word spread, and various friends called and asked if they could also join us. Thus the simple family burial turned into a memorial service, with Don's brother officiating. It was a small group and a simple ceremony, and it felt right."

Sue describes the details that she put into the funeral arrangements. "The funeral arrangements were pretty much left to me, and in our church, there was a lot of support. I was able to get everything together pretty easily. I asked our son to say the family prayer after the viewing, which he was willing to do. I felt impressed about whom to ask for the four speakers for the funeral itself, two of whom were family members. His sister played the piano, and his nieces sang a song Len had written. He had written two songs, and the nieces chose the one they felt good about doing. I felt a spirit of harmony about how the program went together.

"Since we knew several months in advance that his death was coming, I had gotten quite a few photos together, and a friend who lived in another state was willing to put them together into a nice video with music. It was comforting to be able to put this together, and it was a fitting tribute to him. I made copies to give to family members who weren't able to attend, and we had it playing during the viewing both the evening before and the morning of the funeral.

"I chose to do the programs myself that turned out to be 'quite the event' because my printer cartridges failed. Another friend, whom I called, who owns a print shop, volunteered to print the programs as a gift to the family. What a blessing!

"I was very moved by a framed poster the funeral home put together for Len. I hadn't even prompted them, but they came up with such an appropriate one for him to have on display at the viewing and at the funeral.

"It was the photo of Len I submitted for the obituary, and they had put it on a background of a lovely lake and mountain reflection, with his name and birth and death dates and this lovely poem:"

A Golden Heart

God saw you were getting tired
And a cure was not to be,
So He put His arms around you
And whispered, "Come with me."
With tearful eyes we watched you suffer
And saw you fade away,
Although we loved you dearly,
We would not make you stay.
So when we saw you sleeping
So peacefully from pain,
We could not wish you back
To suffer that way again.
A golden heart stopped beating,
Your hardworking hands were at rest,
As God then gently called for
His brightest and His best.

Carrie also took control of her father's funeral arrangements. "After my dad died, I went full throttle into funeral mode. I asked my family if they wanted to help me do any of the work, and I was left to do the tasks alone. I was not going to let Dad down, and I was going to make his funeral a positive celebration of his life. I didn't sleep for several days. I wrote the obituary, made an awesome video of his life, wrote the eulogy, and made the wake and funeral cards. I had to keep busy to stay upright, and that was my way of coping. I interviewed family members and made picture boards, brought in scrapbooks and trophies. My dad had a funeral that would be hard to beat as far as being positive and uplifting. I wanted to make him proud, and I know I did just that. I had to. There wasn't an option in my mind. The best for the best."

Melissa explains her family's challenges. "We had made the decision to have my dad cremated before he died. It was always a given. My mom did not want a memorial service. My brother and I insisted and our church helped to plan it.

"As my dad was sick with the cancer for two years working on and off, money was so much of an issue, so we did everything as inexpensively as we could, and that is what my dad would have wanted."

Aaron remembers what was important to him. "After she passed away, we had an outdoor funeral near the grave site, which was attended by a large number of friends and relatives."

Deborah describes how easy and simple the arrangements were. "The funeral was very easy since my mother had already discussed it with my father. She wanted only her family: my dad, my brother and his family, me, her three sisters, and brother. She had chosen her outfit. She let my father choose the gravesite and basically gave us open reign with the memorial. My family plans on having it at our house once the weather warms up.

"Everyone was kept up-to-date with the situation, so they were able to fly out on short notice. It was a very nice funeral, where each of us got to speak briefly about how much my mother meant to each of us."

Getting through and moving forward

Lisa explains what got her through and gave her comfort during this difficult time. "There were so many beautiful cards, memorial tributes to Fred, food, plants, and calls. I was overwhelmed by the outpouring of support and love. People spent time writing in the cards and checking in with me. I felt surrounded by comfort and caring. I created a memorial Web site for Fred shortly after the memorial service. On it, I added the eulogy, a couple of pictures, a journal with my feelings, and links to the Web sites of Pancreatic Cancer Action Network and the Lustgarten Foundation. I find this Web site so comforting since people can visit it, sign the guestbook, and leave virtual candles and flowers. All of these things helped me feel that I had done everything that I wanted to do to celebrate the special person that Fred was and what he meant to me. I continue to be comforted by the Web site over one year later, and I write in the journal periodically."

"The most helpful tool that I used during this time was working out," explains Ryan. "It helped me free my mind from everything that was going on."

Sherri describes her feelings after the funeral and how she got through this difficult time. "I felt very empty after the funeral and graveside service. Taking care of Mom had been my thing for so long that not doing it anymore just didn't feel right. I was still in shock and wondering had this really happened, what I would do now, and how would I continue on without her.

"During the week after her funeral, when I was off work, I tried to rest and recuperate as much as possible to try to quell some of that exhaustion and return to a seminormal state of being. As difficult as it was taking care of Mom (which I would never go back and change), it was even more so in trying to stop taking care of her and get some rest. In my mind, I wasn't supposed to be resting. I was supposed to be caring for Mom, so I would ask myself why I was just sitting there doing nothing.

"People always say to let them know if you need anything. It's very difficult to determine what you actually need and then to actually ask for it. What probably helped me the most at this point was not having to make decisions. Someone brought food to the house but didn't necessarily ask what I wanted. They made the decisions for me. Just not having to think was what began to restore my sanity (if there ever was such a thing)."

Michele describes her condition after the funeral. "I felt like I was in the twilight zone. I couldn't believe my daddy was gone. The week after my dad passed, the house was buzzing with people, friends, and family, and after the funeral, I just sat there in shock. I was a zombie. I didn't know what to do or how to cope. I was alone in the big house in which I grew up. The best word for my feelings at the time was *desperate*. Still, today, I feel this way sometimes. It is still hard to face the reality.

"The most helpful people at the time were my friends and family, who took over and organized things for me. I was zoned out the whole time and wouldn't have been able to get anything accomplished without them. My friend Lindsey flew up to help me the morning he passed. She ran errands, coordinated things, made me sleep, cleaned up, etc. for the entire week after he died. I don't know what I would have done without her. My uncle helping with the funeral was a blessing too. And of course, my boyfriend, who didn't miss a beat, flew up as soon as possible to be there for me to hug."

Aaron refers to the empty feelings he had after the funeral. "For a few days afterwards, I had an open house for visitors at my home, and it wasn't until the visits were over that I began to realize the full extent of my loss. After living with someone you've loved for fifty-one years, the house felt very empty, and in a strange way, I still expected her to walk in through the front door. Friends and relatives were helpful, but when I went out with them, I had a strong feeling of being lonely even in a crowd."

Judy talks about her need for support. "I had a lot of support from family and friends. I took every offer of dinner, invitations of support, etc. I let our friends take care of me during that time. It was good for me to be on the receiving end of things. Jim and I were always givers, taking care of everyone else. There is a time to be a receiver, and this was the time."

Jeanine found strength in her faith. "As already mentioned, I felt that both events were meant to be, orchestrated by a power beyond me, and they felt right. I felt that Don was honored in a manner that would have pleased him and our Creator, and that is all that matters. I found strength in prayer. 'I am strong in the Lord and the power of His might. I can do all things through Christ, who strengthens me.' A close second to prayer is the support of my family and friends. Without God and the love he has shown me through family and friends, I do not believe I could have survived losing Don . . . nor continue to survive."

Sue agrees with Jeanine. "My faith in life after death and my knowledge of the spirit world was and continues to be a great comfort. The love of family members, both close and extended, and of friends who loved Len and me has continually been a source of strength. I have a painting, which gives me great comfort, of Jesus at the gates of heaven giving a newly arrived spirit a big hug, and an assurance that it represents Len's welcome home. The scripture I selected to be printed on the program gives me great comfort: 'I am encircled about eternally in the arms of his love,' from the Book of Mormon.

"A neighbor I didn't know very well came by a day or two after Len died and offered to help with whatever I needed. I remember feeling overwhelmed by the bag of things that came from the care center, and so I asked her to sort through it and put those things away for me. It was such a tender moment to have someone care enough to do what was needed at the time."

Bev talks about the bereavement group that found her. "A couple of weeks after my husband died, I received a letter from the hospice organization that I had hired expressing condolences for my loss and also inviting me to join an eight-session bereavement group (at no charge) developed specifically for people who had lost a spouse. Initially, I was not interested in even considering such a group, and I threw the letter away. A few weeks later, I received another letter, very similar to the first one that I had discarded. But this time, I was more willing to consider it, so I kept the letter. My husband had died in June, and the group was beginning to meet in mid-September. I decided to attend the first session, telling myself that if I didn't like it, I wouldn't need to return. But I found comfort in the group, which was led by a social worker (who had many years of experience with grief counseling) and co-led by a volunteer. It was helpful that my group was specifically for spouses—and all of the members of my group were women. While there were obvious differences in what each of us had endured, there were also many commonalities. (There was also a separate bereavement group for others who were grieving the loss of a family member, and that group was comprised of adult children and siblings of the deceased.) At each of the eight sessions, there were boxes of tissues on the table, and as we told our stories and listened to the experiences of others in the group, we cried a lot. The group leader followed a curriculum, so each session had an opportunity for a general sharing of information, but it also focused on a particular aspect of dealing with our grief. I felt that the experience helped me, as well as others in the group, to heal. While my group was time-limited, there are also bereavement groups run by other organizations that are ongoing, and anyone can attend at any time."

Carrie remembers her highs and lows. "After the funeral, I went through some very lonely valleys. I tried to hear my dad's words and do as he wanted me to do, but on some days and weeks, it was just hard to function and get through. I knew it wasn't going to be an easy road, as I had experienced grief before. However, I was more determined to continue to make him proud and prove to the world he spent his time well raising me, that his time was not wasted. Everyone gives up some sort of dream to become a parent and makes many sacrifices for his or her children. I didn't want all of that to be in vain or I know he would not have been happy. What helps me today, three years later, is remembering the talks that we had and thinking about the memories that we have made. I feel him in me, and I feel as though I am also living for him. This thought totally motivates me to not waste a single day on negativity in my life, rather to be like him and work and play hard. His 'replacements' were very important to him, and he felt as though he had done his job on earth. I know, without a doubt, that Dad is with us and guiding us each and every day. I talk to him, and I have had some crazy things happen to me that have proven that he is watching over me. For that alone, it makes me get my head out of the muck when I start to sink down into it. I can't waste the life he and my mother have given to me, and I am determined to continue to live and see the world as he taught me to see it. I only hope I can do for my children what he did for me. He taught me how to live, and he, for sure, taught me how to die."

CHAPTER 6

Building a New Life

At the rising of the sun and its setting . . . we will remember you.
At the blowing of the wind and the chill of the winter . . . we will remember you.
At the opening of the buds and the rebirth of the Spring . . . we will remember you.
At the blueness of the skies and in the warmth of the summer . . . we will remember you.
At the rustling of the leaves and the beauty of Autumn . . . we will remember you.
At the beginning of the year and when it ends . . . we will remember you.
For as long as we live, because you were a part of us . . . we will remember you.

—Rabbi Sylvan Kamens

None of us dreamed we would be in this position. We all imagined a long life with our family members. However, that didn't happen. Therefore, we were all faced with how to rebuild our lives. Based on our personalities, prior life experiences, and philosophy of living, we each have done this differently. We have found what brought us comfort during unimaginably difficult circumstances. We also discovered the strength within us, of which we may have never been aware.

Beyond the funeral

Aaron explains, "When the funeral and its aftermath were over, I passed through what probably was my most difficult period of missing her when I was alone at home for the first time. If I misplaced something, my initial reaction was to call out to Frayda to ask her where it might be. It took me quite awhile to realize that it was futile. I found that eating at a restaurant or going to a movie, both alone, bordered on being depressing."

Jackie describes her concern for her niece. "As soon as my sister passed away, the family made the funeral arrangement with her wishes in mind. My biggest concern was her eighteen-year-old daughter who would be living all alone in the house after the funeral. She did not want to be in the house alone, but she also felt that she did not want to leave the house. There was not much money left until the life insurance for her care was distributed. Today, she is in her own apartment and attending her first semester at the local college."

"One of the first things I did after returning home was write a letter to my mother," stated Deborah. "Later on, I ended up putting the letter in her casket for it to be cremated with her. Writing that letter so immediately helped a lot with the healing process. I was able to say many things that I wasn't able to say during her illness.

"Another thing I remember is crying a *lot*. During my mother's sickness, I did my best not to cry in front of her to show her that I was an adult and that I would be just fine without her. I'd always smile in front of her but ended up crying myself to sleep every night. I don't even think my brother or father ever saw me cry. I felt bad about it too since my aunt told me not long before that according to Buddhist tradition, family members should try as hard as they can to refrain from crying since it makes it more difficult for the departed to continue on to their afterlife.

"Two of my mother's sisters flew in the day after her death and stayed with us for a week. Having them around was wonderful: They shared stories, and I was able to learn new information about my mother's childhood. I spent a lot of time pouring through old photo albums and peeking through the lovely memories my mother must have had."

Sherri describes her overwhelming depression. "I didn't really take any immediate actions, with the exception of trying to accept my mother's passing and learn how to go on without her. My beloved fourteen-and-a-half-year-old Pomeranian, Raider, sadly passed away eleven days after Mom did, so it felt like my grieving intensified for a while before it ever started to lighten up. Mom died on February 13, 2009, and our county was gearing up for the Relay for Life event for the American Cancer Society. I was a team captain on a work team and a member on the planning committee, so once I returned to work, I tried to stay as involved as I could, hoping to feel that I was being useful. My uncle Earl, Mom's brother, died on Easter Sunday, April 12, 2009, from cancer. The Relay for Life event was held the first weekend in May. Once that event was over, coupled with that upcoming Mother's Day weekend (my first since Mom died), the letdown was like the day after Christmas. I experienced what I'll call bottoming out.

Michele describes her state of mind immediately after her father's death. "I was in such a daze for so long. I don't think I really had any comfort by sleeping. It was nice to get away from reality by closing my eyes. I did have several dreams of my dad in the days after, so maybe it was nice to see him. Therefore, sleep was my way to get him back. I still dream about him today, but there were a couple of distinct dreams that were so vivid then. I wish I would have written them down."

"After the funeral, I didn't do anything major as recommended in this situation. I continued day to day as before his death, but I spent more time and money eating out. I spent more time with his family and my family," explains Judy.

Jeanine describes the help provided by her family. "My sister arrived the next day and helped us sort through Don's clothes plus get rid of his leftover medications per instructions from hospice. I knew that if I didn't do those things immediately while caring people were around to help me, it would be months or years before I'd be emotionally able to do them."

Carrie explains her life after the funeral. "I honestly don't remember much about the days and months after my dad's death. I only remember going through the motions of being a mom and trying to make it through the days so that my grief wouldn't spill over into the happiness of my two small children. I remember wanting to go to a grief group so that I could talk to someone and be with people who knew how I felt, but there wasn't one in my area. I read some books about grief and loss and kept reminding myself that it was okay to feel sad and, in time, it would get better. I kept repeating the words my dad told me and replaying the talks we had together about what it was going to be like during this time. I wanted to make him proud even in death, and I couldn't let myself get into such a hole that I couldn't get out. I had been there before and didn't want to go anywhere near there. I would hear him say, 'We can't change the cards we're dealt, we have to accept the way things are and not worry about things we cannot change.' Man, he was a lot stronger than we are in the land of the living. I can't seem to let a day go wasted anymore, knowing he is encouraging us to make the most of everything and not waste time."

Just like her daughter, Bette describes her feelings after the loss of her husband. "I told myself I would be a survivor and would not depend on my children. I would use them when I needed to but would try to establish a life of my own.

"They had left our home a few years earlier and had their own lives to live and children to care for. I immediately went to a grief support group that met every Monday evening all year round. This became my new life with new friends. This group split up into a spouse group and had older and more recent losses. I also went to look for a part-time job, which I got immediately in a day care center, which is what I did at home for thirty-eight years. I needed a kid fix, and this was a help to me."

Lisa explains the focus of life after Fred's death. "I had to take care of myself. I had fallen only twenty-four hours before Fred died, and I was in the emergency room. I received thirty-five stitches on my face and had work done in my right eye. During the week after the memorial service, I had doctors' appointments. My daughter and her now fiancé and my son took care of me. I was forced to rest as I could not drive. In another two weeks, we celebrated my birthday. My daughter booked a spa day, and she and I spent a wonderful day together being pampered and relaxing. Despite all of my anxiety about getting through the day, everyone made it special. Then on February 15, I made a decision to change my looks, and I dyed my hair. It was a big decision for me since Fred and I always said that we would keep our hair gray, and it didn't bother us. But now, I felt I needed a new lease on life and a new look. I am very glad I did it as not only do I look younger, but I feel like I am a new person in a different life."

"The first step I took after the death was to assist my mother with anything that needed to be done," explains Ryan.

Finding comfort in groups and family

Aaron describes his search for the right support. "I received a number of invitations to have dinner with friends, but initially, it was with one or two couples and me. Although they all were very friendly, I didn't enjoy being there without the companionship of my wife. Sometimes, a single woman was invited, but still, it wasn't very helpful. A few months after Frayda's passing, I decided I would date some women. Most of them were suggestions supplied by my friends. Some of my acquaintances felt that it was too soon to start dating, but I felt that to hold off possible depression, I would do it. Although I found none of the women I dated to be as appealing as my wife, it did make me feel better to see that, clearly, I was appealing to other women.

"I joined a bereavement discussion group that was associated with hospice in the hospital in which my wife died and found it to be somewhat helpful to see that there were others going through the same period of readjustment. One difficulty, however, was that the group, although constantly changing its composition, was made up mostly of women, the ratio being over 10–1.

"It was about this time that a male friend of mine told me about a male-only bereavement group that had just been formed. The leader was a very bright, a personable man who had lost his wife of fifty-two years to cancer about six months before and was having a great deal of trouble adjusting to her loss.

"I joined this group and found that the discussion of our mutual problems in a male-only environment was much more useful to me than doing it in a mixed group. After about a half-dozen meetings, I felt ready to enter into a real relationship with a woman. It became clear to me that if you've had a happy and close relationship with your wife, it's difficult to picture going through the rest of your life without another close relationship."

Deborah talks about her family relationships. "Spending time with those whom I knew were just as hurt and saddened by my mother's death was the most comforting. Writing the letter to my mother and working on a eulogy that I knew would make her proud helped as well.

"I've always been close to my dad, but he's a very reserved person. He's very methodical and logical and did a wonderful job with the logistics of my mother's illness. I am very close to my mother's sister Irene. I went to school in California, and Irene sort of adopted me as her own while I was out there. I'm not as close with Aunt Wendy and Susanna, but we still spend time together. Uncle David is probably the most distant. My brother and I are not very close since he is so much older than I am, but we talk to each other at least a couple of times a week. One effect of my mother's illness is that we are much closer than we were as children."

Jackie states, "I am most comforted by the fact that my sister is no longer struggling for life. She is at peace. She has two great children that will do well and of whom she would be proud."

Sherri explains her feelings after the funeral. "Despite what I felt, I had the love and support of my family and many friends, including coworkers whom I am fortunate enough to also consider my family and friends. I tried finding reading materials on how to deal with the loss of a parent or a loved one, but nothing seemed to click even though I could relate with whatever information I was reading. I called the local hospice office that had been following Mom's progression and spoke with the bereavement counselor and specifically told her of the bad thoughts I was having. She offered to put me in the next available support group, and that was pretty much all I ever heard from her again. As long and slow a process as it was, I supposed I used myself too as a resource, hoping that my mind would start working and processing things again and would give me some guidance as to how to handle things. I'm not so sure I can honestly say any of this really worked.

"The passing of time helped, along with the love and support from my family and friends. I kept searching and searching for something, I didn't know what, to bring me peace about this situation and to know that I would be okay. Though I didn't recognize it initially, I did start going to church again, and I must say that I am finding some peace there. I started out needing to drive my aunt and uncle to church after my uncle had shoulder surgery and was unable to drive for at least six weeks, but then it turned into regular attendance, and I eventually joined the church on Easter Sunday, April 4, 2010. I think the best long-term comfort I have received is from that. Mom had asked me to promise her that I would get back into church, and I promised her I would . . . I just didn't say when.

While I knew I wanted to get back into attending church regularly for me as well, I guess when I did, I felt as if I was giving Mom what she wanted and had hoped for, and that felt good."

Michele explains her relationship with family members. "Good, actually much stronger than before. We have been making an effort to be a family again, which is wonderful but also saddens me as I wish my dad was here to participate. My cousin has been living in my house (the house I grew up in which was my dad's) and it has been so nice to develop a relationship with him. I know my dad would have liked to know him too, but I cherish the fact that we can spend this time together and may not have otherwise."

Judy describes what she needed after the funeral. "I counted on my family, extended family, and close friends to lean on. I contacted several organizations and went to grief counseling. In particular, I joined a grief group for other widowed people who had lost their spouses to cancer. I began to see my individual therapist again. I had stopped seeing her before his death because of financial considerations. I saw my primary care physician and asked for an antidepressant. He wanted to see me every month for the first six months to monitor my mental and physical condition. I bought several books that were recommended in grief group and by friends. I joined a health club and began to work out regularly, which was recommended by my doctor and several friends for stress relief. I found an online support group and started to meet with local members of that group for monthly dinners. I found that spending time with others who had lost their spouse was very healing."

Jeanine explains the role that religion and family played in bringing her comfort. "My primary resource has been, is now, and will always be prayer and my faith in our Creator. Next comes caring family and friends as resources. When I could feel my emotions spiraling out of control, I would concentrate on intense prayer, and an unexplainable calmness would come over me.

"I am very blessed to have my two children as they are very supportive and caring. My son and his wife want me to move closer to them so I can be a part of their lives and be closer when I need help. My daughter and her family currently live close to me and have been supportive but see value in having me move closer to my son, especially since they also plan to move to the same area in a couple of years.

"My relationship with two of Don's brothers is amiable, and they call occasionally to check on me, as does one of his aunts. I have no contact with the third brother, primarily due to lack of interest on both parts rather than any animosity. My sister hasn't been able to be very supportive lately because of intense problems in her immediate family. However, she was immensely supportive during the first weeks after Don died."

Carrie describes what brings her the most comfort. "The things I feel brought me the most comfort were the things I did with my father. Going to events as we did, taking walks in places that we did, bringing my kids to the same places, and talking about my childhood. For me, I also found comfort at his grave, watering the grass and everyone else's flowers (seriously for hours). I always wanted to leave a note telling other visitors, 'I watered yours. Would you please water mine?' He was a man who made the most of every day and in the most positive way. I am trying to live like him. I feel him motivating me to be more like the person he was. I am trying to look at people and not judge them, accept people for whom they are instead of thinking they should be this way or that way. I try to learn as much as I can because this world has so much to offer us. I feel comfort when I see an eagle. I feel comfort in a sunset, a sunrise, and any other thing in which I see beauty. As I always have said, I had no regrets when my dad died because I did so many wonderful things with him. I got to experience more with my dad than most people, and I have chapters of memories stored in my mind and heart. I can dig them out whenever I need to, and most of the time, I smile now. My goal is to make those kind of memories with my children and to live my life in a way that would be productive and make a difference in the lives of others. My dad would always say at his time of death that if I become a sad, depressed, and cold person after he died, it would be such a waste and his accomplishments and work with his children would be in vain. But if I could dig deep, lift my chin up, and carry on as if he were walking next to me cheering me on, he would be so proud. I want to do that. It is not always easy. I believe that my dad would be happy if he knows his family and loved ones are safe and happy. No matter what it took to get them to that place of being happy. Life is for the living, and even though my world went colorless when he died, the color is coming back, and I am appreciating every moment of it."

Bette states, "I got active with Pancreatic Cancer Action Network and helped with the bike ride and also stayed with the LISTSERV to offer what I had learned and had been through to others. I also learned that I was not the only one and that they were trying to raise funds for research for future years.

"I got a lot of comfort from my support group as they all had walked the walk and understood my feelings of sadness and happiness. I did get some comfort from family and children but not as much as I thought I would. I realized that my family was not as close as I thought we were. I felt uncomfortable doing family things that we had done before. I actually avoided going to family functions and family members' homes.

"I was running and tried to stay busy all the time. If I had nothing to do, I would go to stores or shopping centers and walk to be around people. I saw a lot of my new friends in my bereavement group and organized plans so we could be together. I felt most comfortable with them. I hated to come home to an empty quiet house."

Lisa describes the role of support groups and family to bring her comfort during this difficult time. "I found a bereavement group through a local hospital, and in March, I began an eight-week program for those who have lost spouses. At the time, it was what I needed—to be with others who really understood how I felt and the effect on my life. I am still friends with some of the participants.

"What brought me the most comfort was hanging on to memories. I decided to create a memorial Web site for Fred. When I was satisfied with all that I had created, I sent it out to everyone that I felt would like to see it—family, friends, colleagues. Each time they would visit the site and leave a flower, a candle or sign the guestbook, I was notified. This gave me great comfort that people had not forgotten Fred and that he still lived in the memories we all had of him.

"In addition, family and friends continued to call me to find out how I was doing. As time went on, they called less, but everyone goes back to his or her own lives. It was also comforting to have people still living with me in my house. In the same year, I lost my two Labrador retrievers and my oldest cat. Therefore, the house would have seemed very empty if not for the three boys I had living with me. My son was still in the house, and the guy we took in after Sari, my daughter, had left for California in 2008. I also took in another friend of the guys. Therefore, with three guys in the house, there were always young adults in and out of the house. I liked the busyness of the house, which made me feel a little less lonely.

"I had decided that to get through the first year, I would do whatever helped me feel better. Therefore, for what would have been our thirty-first wedding anniversary, I went back to the place where we were married and celebrated our twenty-fifth wedding anniversary with my children and some friends. We had a lovely dinner, and I felt that Fred was right there with us.

"For Father's Day, my children and I went to Syracuse. It was my wish that the ashes be spread in the falls where Fred and I used to go when we attended Syracuse University. It was very emotional as we also took the ashes of one of our dogs that had died on May 8. I felt like it was the right thing to do.

"In July, I went to visit my daughter and her fiancé in California. I decided that flying is okay and that I would not allow my fear to stop me from traveling anymore. In October, I went alone to visit my friends who own an inn in Vermont. I did the drive alone and was very proud of myself. Finally, in December, when we would have celebrated the winter solstice, I went out to eastern Long Island and treated myself to a spa day and overnight stay at a resort. The snow came down the next day, the day before the winter solstice, and since Fred and I always walked in the first snow, I felt that it was a sign from him. Finally, going to my job and knowing that I had this identity has continued to give my life meaning and stability. All of these things brought me comfort as I got through the first year without the love of my life."

"What brought me the most comfort was spending time with my mom," Ryan states. "We spent time just talking about the future and what we were going to make of it. I know she loved hearing about what ideas and what plans I had even though it was such a hard time. It was good to think about the future and not just dwell on the past."

Sue describes her life after the funeral. "I spent some time at the Brigham Young University library, making a CD from the cassette recording that was made of Len's funeral, which I shared with many family members and friends. I made copies of the slideshow I had prepared for these same people. I remember sorting through a lot of papers and dealing with the financial and legal aspects of my husband's death. I experienced more freedom not having to be concerned about his care. I took a trip to Los Angeles for a few days, which I thoroughly enjoyed. One friend I visited had quite a few books on grieving, and I selected five of them to take home to read and then return to her."

The challenge of change and moving ahead with life

Aaron explains the process of dating. "About six months after my wife's passing, I was introduced to a woman who had lost her husband after many years of a happy marriage. We have been dating for about eight months and find we enjoy each other's company a great deal. Because of my successful experience in the male bereavement group, I decided to join its board of directors and help to expand the organization to other locations, even though personally I feel that I do not need the same assistance as I did before.

"Life is better now, and although I still think about my wife frequently and miss her a great deal, the ache in my heart definitely has subsided.

"I'm not what you'd call a religious person, but I feel there might be some invisible guidance in this: About six months before my wife, Frayda, was diagnosed as having stage IV pancreatic cancer, at a time when she had no hint of even having a problem, my oldest son Jeff formed a biotech company, American Gene Technologies. Its corporate mission was, and still is, to develop a cure for cancer based on the use of genetic cell therapy. What is even more interesting, the company chose as its initial goal the development of a cure for pancreatic cancer. During the period when Frayda was fighting the disease, the company made large strides in the direction of formulating and testing a cure. The basic technique involves using a viral vector to locate all of the diseased cells and then modifying the cancer cells' DNA to prevent the cells from growing or reproducing. To date, the results have been excellent, and some large drug companies currently are looking into providing funds for more extensive testing. Hopefully, future results will show that what has been extremely effective in live-mouse models that have been given the disease will be as effective in the cure of humans."

Jackie describes the changes in her life. "For me, life has slowly returned to taking care of things in my life. I know it won't be the same because many things have changed for me. I rejoined the gym and am working on getting rid of the fifteen extra pounds that I gained sitting in hospitals. I had been in a new relationship, and he has been wonderful throughout the process. Today, I can say he is a keeper. I completed my first college degree and have a full scholarship to finish my four-year degree and then on to get my masters. I finished fifty-eighth in the country in my real estate business this year. I am working hard and playing hard. I have been to Alaska and will depart for Italy and Norway in a few weeks. I am living life full throttle, and I don't want to miss a thing. I feel that I've made good things come from bad. My sister's illness made me know, understand, and appreciate how short life is and to make each day count."

Deborah describes her focus now. "Since my mother's death, I've been spending a lot more time with my father. I live in New York City but go back to spend almost every weekend with him, cooking and cleaning, running errands, etc. It's not so much an obligation as a desire, although I have to admit that I do miss making time for myself.

"I thought I would be hearing more from my mom's sisters, and I'll admit that I am a little hurt that I've heard nothing, even during the holidays and Qing Ming Jie (a Chinese holiday, which involves visiting and cleaning the graves of loved ones). My cousin wrote me an e-mail recently, saying that her mother is still so hurt and saddened by losing her sister that she is still in a kind of denial. I've realized that my immediate family members are not the only ones drastically affected. I knew my mom for twenty-seven years, my brother knew her for thirty-four years, my father for thirty-five, but her sisters and brother knew her for forty, fifty, even sixty-something years.

"I don't buy a Mother's Day gift anymore (or a birthday/Christmas gift for her). Something in my throat swells up when I go to a bridal shower hosted by the bride's mother. I say 'my dad' instead of 'my parents,' and I don't spend any more balmy spring days wandering around Central Park with my mother.

"My mother passed about seven months ago, and I'm still helping my father to clean up the house. We've donated a lot of her clothes. My father sold some items, and now that the weather is warming up, we've been able to spend time decorating her grave and fixing up her garden.

"I think I was able to move on a little earlier than most because I had prepared myself during my mother's illness. It still is absolutely overwhelming, but I was able to piece myself together not too long after her death. I still think about her every day, but it's only on the very rare occasion when I break down and cry.

"I still keep on top of breaking news about pancreatic cancer, and I'm very involved with Pancreatic Cancer Action Network. In fact, I recently took lead of the Manhattan affiliate's fund-raising effort for the Hamptons Marathon.

"I recently decided to apply to business school for several reasons: Eventually, I will need to settle down in the suburbs, as opposed to a city (where most of my current career options are), in order to take care of my father. Getting an advanced degree will give me more options at smaller, regional companies. I want a last hurrah, the ability to live and study in another area, before returning back to New Jersey/New York. If I want to go into the nonprofit industry, having an advanced degree is very useful. Ongoing goals for me include continuing to raise awareness and funding that may one day find a cure."

Sherri describes the drastic changes in her life. "Let me preface this with saying that Mom and I lived together for the last eight years before she died. We were both divorced, and I have no children (except the four-legged furry kind). What didn't change in my life? I no longer could walk down to her room in the middle of the night when I couldn't sleep and find her not sleeping either, and we'd just sit there and talk . . . or not talk, which was okay too. Who would I call when I had something big to tell somebody that couldn't be repeated because I knew Mom would keep my confidences? Who would let me whine and have a pity party when something didn't go right or when I was sick? Who would scratch my back when I had an itch I just couldn't reach? Who would get the phone or the door when I was in the bathroom? Who could I call on to bring toilet paper to the bathroom when I realized only too late I needed some (yes, it's the little things)? Who would unconditionally love me no matter how bad a mood I was in? Who or what would give me a reason to go on in this life? Who would be proud of me for any little thing I did? I just have to deal with these issues as they arise and do what I can, as best I can."

Michele speaks about her new life. "My boyfriend moved from his hometown to live with me in my dad's house while I sort through my father's business dealings. It is wonderful and challenging, and I don't know what I would do without him. He is an amazing, supportive, and loving man, and I am so thankful that I have such a great love in my life.

"Every day I now talk to lawyers, accountants, tenants, and my dad's bookkeeper. My father was in the real estate business, and it's a tough market in which to be, so I am not only getting a quick education on the industry but a quick education on running a business in the toughest of times. It is very frightening, but also I am proud to try to keep things running the best they can. I have had many struggles in this process, and I feel like I am learning a life lesson every corner I turn. Some are good lessons, some are so hard, and some I plain wish I didn't have to learn at age thirty, but I guess it will only make me stronger."

Sue speaks about her adjustment to her new life. "Our hospice provider had a support group for us, which I attended each week for the few months after Len died. I was kind of bouncy. Most of my sharing was upbeat, and I reported on the many activities in which I was involved and the sense of freedom and relief I felt after the many months of my husband's suffering from pancreatic cancer. The group leader used me as an example of what people 'should do'—like go to the gym, keep working, look for the positive, and enjoy nature and music. I see, now that four years have passed, how cavalier my attitude was and how unlike the others I was—but as I look back, I realize more than ever how important it is to acknowledge that each person's grieving is individual and we each have a timetable that doesn't really compare with another's.

"I did a lot of reading and watched movies that I thought might make me cry. I listened to a particular address by Rex and Janet Lee (former Brigham Young University president and his wife) where they shared very personal feelings and experiences as Rex went through the stages of cancer and remission and then asked to be released from his responsibilities as president of the university just six months before his death in March of 1996. Listening to this talk invariably brought not only tears but a flood of emotion and sobbing."

Judy explains the pain of everyday life. "I wake up each day with the knowledge that I have lost the most important person in my life. I cry every day but usually for a short time. I look forward to the future, but I know that I will never feel his love, the touch of his hand, his kiss ever again on this earth. I live as well as I can, but I have aged, and I am sad much of the time.

"I am trying to look forward and not dwell on looking backwards. I hold Jim in my heart, but I need to live on and make a living. At first, I couldn't work as much as I should have. As time goes on, I find more motivation to go forward. I honor his memory at significant times."

Jeanine describes the changes in her life and a metaphor with plants. "Virtually everything in my life has changed, other than the basics (i.e., breathing and other basic bodily functions). When Don was alive, my life was all about him . . . even before he got sick. My son shared the following theory with me shortly after Don died, and I think he eloquently explained my situation.

"The plant metaphor is a theory on relationships. It goes like this: I see when two people come together and start to interact, it is like taking two plants being planted in the same pot side by side. There is a metaphor about plants and their interdependence that my son described. The plants appear to be separate individuals just aligned and matched by proximity. What the outside world doesn't see is what is below—the roots, the heartstrings, the attachments we have emotionally and spiritually in our lives. The outside world can't see these roots, but the interaction of the roots shows up on the portion that the outside world can see, and it shows up in the health of each plant/person.

"Another thing to take from the metaphor is what happens to each plant when the other plant 'leaves.' No matter how it leaves (is removed, leaves on its own, or dies), how it affects the other plant/person is dependent on the roots' structure. There are no deeper ties below the surface of these two plants/people, and if one leaves, the other just keeps on going. We've all seen couples like this. There is also the case that what happens depends on what plant leaves. If the plant that is sharing all its roots leaves, the other plant will keep going since it has retained and kept most of its roots. If the one that kept all its roots leaves, then it takes nearly all the roots of the plant that shares with it, for you cannot separate intertwined roots that grow together. And if it takes all the others' roots, it will wilt and be severely damaged. That's an unbalanced relationship, and one will suffer greater than the other upon separation.

"With most of my 'roots' ripped away, I went from doing everything with Don to having to do everything without him. That impacted nearly every aspect of my life."

Carrie describes her philosophy of life. "When you experience death, at least for me, you realize the finality of it. You see how fragile life is, and you, hopefully, make the changes you need to make to ensure that you don't waste precious time on unproductive matters. It's not always easily done, but death makes you grow and learn about yourself in a way you never thought existed. If it's a death of someone of great importance to you, it puts you on a tightrope, and you sway back and forth, just trying to hang on. I do believe it depends on the timing of the death and where that person is in his or her life. But to accomplish coming back to life for the ones left behind and becoming a better person because of it helps make the sting of losing your person a little less in vain. You can't change the outcome, but you can live again . . . and you sure do appreciate the good times of both yesterday and today."

Bette shares the changes in her life. "I have grown and gone on with my life. My husband left a journal to me, saying he wanted for me to go on and live a life for me—find someone to love me and take care of me and with whom to do different things. I wanted 'me' to do the things I always wanted to do and did not do them with him as I was a pleaser to him.

"I did do that and am in a relationship with a fellow I met in the bereavement group that also lost his spouse the same time as I did. We are both moving on and have a wonderful time together. The feelings will always be there for my husband. I miss him every day and am constantly talking about him and enjoying the memories. I have found that you can love again, and it is in a different way—but also a good way. My goals are to be happy and to start a new life with my new friend. Will I marry again? Time will tell, but neither of us has any plans right now. Life is for the living, and I am finally getting it. I still have days when I wonder what or where I belong and if I am doing right, but they are further apart than when I started. This is what my husband of almost fifty years (would have been on September 17 this year) would want for me—to enjoy grandkids and go on with my life."

Ryan describes the changes in his life. "I was now the man of the house. I could not rely on anyone else to do anything. I had to pretty much step into my father's shoes and figure things out for myself, regardless of what it was. I felt more stressed as time moved on. It really hit me that I would never see him again and the burden of maintaining the house and looking out for my mother was on me. One of my goals for the future is to go to school for construction management to become a project manager. I hope one day to partner with somebody and own my own construction company."

Lisa explains how she has dealt with the changes in her life. "The biggest change in my life is to live life now as a single person. When you are part of a couple, you not only spend time together, but you think as a couple. Now my decisions are all my own. I found a young widow/widowers group that gets together and shares activities. I joined a singles group that offers other opportunities for fun. I have been meeting lots of new people, especially women. I try to focus on what helps me get through the loneliness and do things that are right for me. I avoid situations that will make me feel uncomfortable or will create sadness. I definitely cannot go to dances, as Fred and I were excellent dancers. I don't and won't go to bars. I have been meeting more and more friends for dinners. I have no interest in dating or attending situations to meet men.

"I am trying to concentrate on taking good care of myself. I was so busy taking care of Fred that I definitely neglected my overall health. I am seeing new doctors and having tests done that should have been done a couple of years ago. I also am finding out more and more about the things in which I find enjoyment and what makes me happy. It's so different just thinking about me.

"As the time without Fred has increased and now that I am in my second year, I am more comfortable with some changes. The guys in my house bought me new dishes and glassware for my birthday, and I was okay with donating my original pieces, which were from our engagement and wedding. The guys worked on changing a room in my house into a new dining room, and I moved my office to the den. I will continue to make other changes in the house, including changing pictures from our trips together to ones that I have taken by myself. I feel like the big challenge of getting through the first year is over. Therefore, this second year is the time for me to explore new things and be more open for change."

Advice from the Battlefield

We have taken you through our journey. We have shared with you our family lives before cancer and through all the stages. As we look back over the course of the years or months of this disease, you have become acquainted with our family members and with the patients. We are now ready to offer our personal advice based on our experiences. There are no rights or wrongs in these writings. They demonstrate only what each of us has learned and would like to share with you in an attempt to offer you support and the knowledge that you are not alone in this journey.

The Survivors and Their Family Members

Pat

Update on Charles's Condition

As of June 5, 2010

"As of now—almost seventeen months after diagnosis and sixteen months after finding the tumor was inoperable and life expectancy was, at most, six to nine months—Charles has no evidence of disease. No signs of metastases have ever been found. His CA 19–9 markers have been normal and quite steady for four months now, and PET scan seven months ago was clean. We are waiting for new CT and PET scans next month and will then discontinue chemo, assuming both are clean."

Despite all this great news, he is very short of breath, lacks energy, and really lacks stamina. For the past two months, doctors have been tinkering with medications, hoping to find a magic combination. We are so hoping that getting off the chemo will give him a chance to recover and live a relatively normal life for however much longer he has, whether it is only a year or two, or an actual cure. For those just starting this journey, there are the occasional good outcomes.

Advice

"Simply put, people who may have or have just learned they have pancreatic cancer (or any other serious disease, for that matter) must immediately take charge of their own medical care, and they must have a family member, friend, or professional to act as an advocate for them. Do not ever assume that any appointment will be made by someone else or that anyone involved in the patient's care will do what they should do when they should do it or that anyone else but the patient or his advocate can ever ensure that the patient receives the necessary care."

Kathy

Update on Vinny's Health

"Looking back, speaking to family members of those who had pancreatic cancer personally helped me a great deal. I had two friends who were wives of the patients. Their spouses had been diagnosed one to three years earlier than my husband, Vincent. Both of these women were generous with their time and their compassion in explaining things and providing an ear when I was overwhelmed. Some of the most practical advice came from those who had walked the same path.

"Early on, in the first three to four months of the diagnosis, I spent a great deal of time researching and educating myself so I could both understand what the doctors said and ask important questions. One warning on research, however: the Internet can have false and upsetting information. As I researched, I was careful to use trusted and reliable sites and to write out questions to check with the oncologist about. On the other hand, the Internet can be an invaluable resource if used appropriately. I also was sensitive to what my body told me. In other words, I paid attention to when to back off because of feelings of fatigue and overkill. I learned there is such a thing as 'too much information.' In addition, 'a little knowledge can be a dangerous thing.'

"I learned that the kinder I was to people who were harried in their jobs, the more I tried to put myself in their places and the more their treatment of me improved. In my case, trying to give something back to an organization that helped us made me feel better. About a year into my husband's treatment, I actively started to investigate what I might be able to contribute as a volunteer. I did this slowly and carefully, being mindful not to overextend myself. And a word on fatigue: for the first part of my husband's illness and treatment, I was so fixated on him I would not have noticed if I had a broken leg. It took many times of hearing (over a period of months) 'You have to take care of yourself too' before I finally began to hear it. I would make a plan to do something for myself and then find endless excuses not to go through with it. However, ultimately, I experienced the obvious: taking care of myself helped my resilience for taking care of my husband.

"I was lucky enough to be given some advice that I actually was able to hear:

> Learn from yesterday, live for today, hope for tomorrow.
> Everyone is different and responds to treatment differently.
> Therefore, time frames are not particularly helpful, except perhaps as
> a very vague guide.
> I learned not to victimize myself by comparing tenths of a centimeter
> in terms of tumor size. Our oncologist would say to us, 'It also matters
> how the person looks and feels.']

"One day, I heard an interview with the wife of Randy Pausch about a year before he died. She was asked what she did about terribly frightening and recurring thoughts. She said she repeated to herself the words 'These thoughts are not helpful' and tried to move away from them. I remember thinking, *Yeah, sure,* and yet when my own husband was diagnosed coincidentally about a month later, her words came back to help me, and I did find them helpful.

"Lastly, I learned (in part) it is a very tough struggle to learn how to let go of (the illusion of) control. We don't have final control. We can only do our very best to love and care for our beloved.

"As of this writing, Vincent has been without treatment since June 2009, and his tumor is stable. We are facing a scan in the next twenty-four hours, with a report to follow in four days. Memorial Day is upon us, and I thought it will be hard to wait for results. We will drink coffee together in the morning, hold hands at some point during the day, and be eternally grateful for the time we have been granted."

Jenifer M.

Health Update

"From 12/08-12/09, I was on oral Tarceva only. I have felt very well, traveling a lot and enjoying every day. I had monthly port flushes along with CA19–9's done at that time. PET scans were every three months.

"On 12/09, a 7 mm lung nodule had grown from 2 mm the previous two months, and my CA 19–9 had slowly risen to fifty-nine. It was decided that I would do three CyberKnife treatments. I did well with that after a liver metastasis in 2/2008. I did that again with no side effects.

"At this time, 6/7/10, my CA 19–9 is back down to nine, and I am still living every day as best I can. I am on no preventative chemo or treatment at all, and that is a bit scary. You could say I live month to month, but if I really put it in my head that way, it would not be possible to enjoy life.

"My advice to anyone living with pancreatic cancer would be to have a positive attitude above all else. Being pessimistic and mad at the world does no good for anyone. Secondly, you really need a grounded support system. I have my children, my family, and especially my good friends to lean on. If I get depressed, I only allow myself one day to wallow in it if I feel the need. But any day wallowing in self-pity is a day wasted.

"I have been very lucky in this journey, being diagnosed three years and nine months ago. All treatments and surgery have worked so far. I am not a diabetic and can eat whatever I want. There were some terrible days during treatment, but it was well worth the last year and a half that I have had feeling great.

"My best friend, Debbie, gave me this card last week. It sort of sums up what I want to say:

What Cancer Cannot Do

Cancer is so limited.
It cannot cripple Love
It cannot shatter Hope
It cannot corrode Faith
It cannot destroy Peace
It cannot kill Friendship
It cannot suppress Memories
It cannot silence Courage
It cannot invade the Soul
It cannot steal eternal Life
It cannot conquer the Spirit.

"Love and prayers to all who read this."

Joe

Health Update

"Healthwise, I've been fine. I have gotten very involved with the Pancreatic Cancer Action Network in Baton Rouge and New Orleans, and I am on the survivors network of Pancreatic Cancer Action Network, advising other survivors who call me from all over the nation.

"The cancer is never far from my mind. I think of it every day. However, I've come to terms with my mortality. I think all cancer survivors do. You don't want to, but you have to."

The Family Members Moving on Without Their Loved Ones

Judy

"The diagnosis of pancreatic cancer was devastating, but we were fortunate that Jim had an aggressive GI doc who believed in second opinions.

"I will reiterate one important story from Jim's journey of pancreatic cancer. I have told this story many times, for the lesson is this: be aggressive in seeking a second and third opinion! Don't let the first doctor tell you to 'go home and die' without at least checking around first!

"When Jim had his FNA (fine needle aspiration) and biopsy of the sample to verify the suspicion of pancreatic cancer, he had *two* GI docs do the procedure side by side in a procedure room at the hospital, which is a well-known and respected teaching hospital in suburban Chicago.

"Dr. A came out to talk to me after the procedure. He said, 'It's definitely pancreatic cancer. There's nothing we can do. Take him home, and get his affairs in order.'

"I cried and asked, 'What about chemo or surgery or other treatments?' Dr. A responded again, 'There's nothing that we can do,' and he said that 'Dr. B will also be out to talk to you in twenty minutes.'

"Well, I sat there—totally depressed and heartbroken—and cried, thinking that Jim would die immediately or in the next week or something.

"Dr. B came out twenty minutes later. He said, 'It's definitely pancreatic cancer, but I want to talk to some surgeons. My office will call you.'

"Dr. B's nurse called the next day, and we had a second opinion with a very renowned pancreatic-cancer surgeon a couple of weeks later. Jim did have to undergo more testing to see if his tumor was resectable. Jim had successful Whipple surgery in December of 2005, had chemo and radiation, and was in remission for nearly two years. After the diagnosis of recurrence, Jim responded well to the Gemzar and Tarceva for ten months. During that time, he worked full time, and we were able to celebrate our thirtieth wedding anniversary with family and friends. We were very grateful for the three-and-a-half years that we had together after his diagnosis.

"I wonder how different things would be if he were first diagnosed now, in 2010, instead of 2005. There are continually new things being learned about pancreatic cancer, and new treatments may offer more hope and maybe a cure."

Beverly

"Learn from all available resources:

"The Pancreatic Cancer Action Network (www.pancan.org) was an invaluable resource that helped us numerous times. My husband and I were inspired by attendance at two of Pancreatic Cancer Action Network's fabulous free conferences, and their counselors provided me with great information via e-mail and phone calls.

"The Association of Cancer Online Resources (ACOR) has a free LISTSERV focusing on each type of cancer. To subscribe to the pancreatic cancer list, go to www.acor.org/pancreas-onc.html.

"When my husband was diagnosed, I purchased an ordinary spiral notebook, which I carried with me to every doctor's appointment. In that notebook, I kept track of all my husband's health issues, doctor's recommendations, chemo protocols and side effects, as well as an ongoing list of questions for the oncologist, nutritionist, and other specialists.

"I also logged in suggestions that I learned about from other families.

"Make connections with other caregivers who have a loved one with pancreatic cancer. This can be done in person (families who are having chemo at the same site as your loved one) and/or online.

"Reach out to friends or relatives to help you when times are difficult. Friends often say that they want to help, but they don't know what assistance you need. It helps to make a list of the types of things you need help with (e.g., picking up items at the grocery store or pharmacy), and when friends ask what they can do to help, you can let them know what assistance you need. There are times when it just isn't possible to do everything yourself, and you will need help from your friends or other relatives.

"Plan outings—day trips or vacations—for days when your loved one is likely to be feeling okay. After a while, you will know how your loved one reacts to chemo, and you can predict which days they are likely to be feeling well enough for enjoyable outings. After his diagnosis, my husband and I were able to take three vacations, which were a great source of enjoyment for both of us.

"Suggestions for the caregivers:

1. Carve out some time for yourself. For example, I continued my membership at the local gym, and there were many days that my husband and I went to the gym together to walk on the treadmill. We also took walks in the neighborhood, weather permitting. Other times, I went to the gym by myself. Caregivers need to be as physically and emotionally fit as possible in order to provide the best care for their loved one.
2. If the caregiver is employed, she should keep the job even if she needs to take a temporary family leave. Being at my job provided a helpful respite from the daily worries and stresses of my husband's health status, in addition to the value of having the additional income. I was fortunate that I could take time off from work as needed.

"Ongoing advocacy and involvement is needed with organizations (e.g., Pancreatic Cancer Action Network, Lustgarten Foundation) that do fund-raising aimed at all aspects of pancreatic cancer: prevention, early detection, and finding a cure for advanced cases of this cancer. The National Cancer Institute should be devoting additional financial resources to PC as well. Sadly, when loved ones die from pancreatic cancer, the survivors are often so exhausted and grief-stricken that they shy away from continuing involvement with the organizations that are dedicated to this vitally important cause!"

Jeanine

"The first advice I'd give anyone going through this disease, or any troubling situation, is to pray and look to scripture for guidance and strength. That is what Don and I did, and I firmly believe that is how he was able to survive two years beyond what was expected, with much of that time being enjoyable.

"I don't know if the next advice I would give to anyone newly diagnosed with pancreatic cancer. However, it is possible to follow in most cases.

"Looking back at our circumstances (primarily the pain Don was experiencing) in early 2006, I'm uncertain if we could have followed my own advice. But for what it is worth, I would advise anyone who has been told they may have pancreatic cancer to immediately research all treatment possibilities, especially alternative treatments, before starting radiation and/or chemotherapy. Once radiation and chemotherapy do their damage, the alternative treatments have a tougher time of being effective. However, regardless of when allopathic treatments are started, a change in lifestyle and embracing alternative treatments is always beneficial.

"Since we did immediately put our trust in our Lord to guide us in all decisions, I now trust that the course of our journey was as God directed it. Thus, I can't say that I would have done things differently if I could. Sometimes, I wonder if Don and I should have talked more about the possibility of his dying and my having to live without him, but neither of us was capable of doing that. Thus, I realize that I must trust that it was best that we didn't.

"Perhaps the best advice I can offer comes from Don. On June 29, 2009, Don posted the following words to his CaringBridge site:

> Although I have been a Christian since age eight, in some ways I hadn't advanced very far in terms of living the "full" Christian life. There are blessings that I've missed out on. One of the things that have been added to our lives since this cancer ordeal is praying together. I don't mean giving thanks to God at mealtimes or praying in small, intimate Bible study meetings. What I mean by praying together is

this: at least once (or more) times a day (usually at bedtime), we hold hands and unload everything before God. We expose all that is on our minds. That all includes thanking him for blessings for each other, asking for forgiveness of any and all things, and asking for help for ourselves and for others in all matters.

We had never done joint prayer before my illness, and we now know we have missed out on such a blessing in life. A lot of mistakes, hurts, and misunderstandings could have been avoided through the years, and also, I believe God would have blessed us so much more.

I encourage everyone who believes in God and Jesus the Son to engage in joint prayer with your spouse, if you don't already. I know some of you are much faster learners than I. I'm sure Jeanine would have been agreeable to this intimate relationship, but I was the introverted, silent type in my spiritual life, so I never suggested we do this.

Praying together may be awkward at first, but it gets easier very quickly. When you're talking before God, nothing is held back. The truth comes forward. Be prepared for raw emotions to arise. I was never one to want to cry in front of anyone . . . not even Jeanine. But crying we have both done, and it definitely bonds us to each other and to God. We understand each other much better, and our love has deepened. We also know each other's commitment and trust in God and how we feel about each other.

I may sound like Dad from years ago with my talk of God and prayer, but I am convinced of its value. I was a constant witness to how God blessed my dad in his dedication and boldness with his walk with the Lord. So if I remind myself of my dad nowadays, that is a good thing.

Each night, we go to bed with complete peace, and each new day brings hope. Whatever happens, we can say that some of these days have included the best moments of our lives, although they have ironically occurred during the worst time of our lives."

Bette

"Remembering all the memories Floyd and I made together during our forty-seven years of marriage and four grown children is what he asked me to do and to live on.

"After he was diagnosed with pancreatic cancer in July of 2006, he decided he wanted quality of life and to make more memories. Seeing family and friends was important to Floyd. He called it Living the Dream. Fortunately, he was able to be up and about going to the cabin, on a Panama Canal cruise, and trips in our new motor home that we had just got right before he was diagnosed. He tried to keep his life normal and even went snowmobiling a week before his passing.

"He had a positive attitude and told us all this is not what he wanted but it was the cards he was dealt.

"He asked us all not to feel sorry for him but to enjoy the time he had left and to be happy. He put on a good face for us, never showing how he really felt at times. He never complained—just took each step of the way, thinking it was one more day he was here. I had to support that request as hard as it was. I never knew what the next day had in store for us.

"He insisted on seeing and calling friends he had not seen in a long time, and we went places for him to say his good-byes. He loved racing, and we were fortunate to travel to a lot of racetracks. Some we went back to before his dying, and he asked for time alone and would walk around by himself, saying his farewell.

"I feel I was cheated out of our retirement as we both had only one short year of it, with many plans ahead for the coming year. We were selling our home and moving to the cabin and traveling in our motor home during the winter.

"My primary role in caring for Floyd was to support him even when I did not agree. Our family did a lot of research so that we felt we did all we could for him. He was treated at a well-known hospital and did Gemzar, which helped for a few extra months. He then did a Rexin G trial that had no effect. We stayed at the hospital for three weeks, with treatment daily for six hours, including weekends. I was at his side all of the time.

"He died on January 11, 2007, living six months and four days with pancreatic cancer—stage IV with metastasis to the liver.

"My advice is to let the patient be comfortable making his/her own decisions and support whatever his/her choice is. Have him/her write in a journal as often as he/she will. Floyd did, and it helped me a lot in my recovering.

"He told me to go on with my life and to do the things I always wanted to do. He also hoped I would find someone to care for me and enjoy me as much as he did. He said, 'I am dying, not you, and life is for the living. You must go on.'

"I have taken his advice. I miss him every day and think of the things we did and were going to do. This September 17 would have been our fiftieth wedding anniversary, and we were so looking forward to celebrating in a special way.

"In my new life, I will celebrate and I do remember all our special days. This is what keeps me going. I cherish the life we had together for so long. I live with special words Floyd said before dying: 'I made some good replacements' and 'Work hard and play hard.'

"We did that. It's just that his life was too short at sixty-six years old."

Melissa

"Spend all the time that you can together. Have no regrets.

"Ask doctors questions if you don't understand. Don't feel like you are being a pain. They are being paid by you to help you or your loved one.

"Reach out for support through Pancreatic Cancer Action Network and sign up on the LISTSERV http://www.acor.org/pancreas-onc.html. These folks have been invaluable to my family and father. They will support you every step of the way.

"Family members, even though it's hard, try not to remind your loved one every day that they are sick. I know I would ask my dad every day, "How do you feel?" One day, he finally looked at me and said he didn't want me to ask him that every day. Some days he just wanted to put the cancer behind him, and it was a constant reminder to him that he had the cancer.

"Family members . . . advocate, advocate, advocate. Be a constant advocate for your loved one. They need you now more than ever.

"Never give up!"

Jackie

"Looking in the rearview mirror from the time of the diagnosis of cancer, there are a few things that I would have done differently. I would have made my sister more comfortable. She did not like anyone taking care of her and insisted on doing things herself, and so we let her. I think she really wanted a lot more help than she would allow, and if I had to do it over, I would have just done those things that I wanted to. It was simple things, like cleaning out her garage so she could park her car inside. After her passing, it got cleaned out, and we all cried because it was something she always talked about but would not let anyone do.

"Today, I also know how lonely she was throughout the illness. She would tell everyone that she was busy and had things to get done. We honored her wishes and left her alone. Later, we learned that she lay in bed alone those days, sometimes not able to get herself even a drink. If I could do it over again, I would not accept her request to be left alone for days on end. I would go and be with her since I now know why she was saying those things.

"My advice for anyone experiencing a terminal illness of a loved one is to be compassionate and very patient. More importantly, don't forget your own needs, and make sure to take care of yourself. You can't fix it, but you can make it easier for the patient. Just don't lose yourself in the process. Share the load—no one can do it alone for too long. Make something good come out of this experience. I started college at the age of fifty, while my sister was ill. I took mostly online and a few night classes. Two-and-a-half months after my sister passed away, I fulfilled one of my lifelong dreams. I graduated with a 4.0 average, thanks to my sister's inspiration."

Carrie

"My advice is to surround yourself with people who care about you. Surround yourself with people who support how you feel. Do what you need to do to survive, no matter how crazy others might think you are. No one knows how it is unless they have been in your shoes, fighting the fight to do anything and everything to make your loved one get well again. If someone makes you feel like you are crazy or wrong for not accepting the illness or wanting to do more, don't listen. You don't want to have regrets, and you need your own space and time to work through the acceptance of the illness. My last and most important piece of advice is don't let your person who has the cancer take care of your emotional needs too much. I was so distraught and unbelievably sad that I was going to lose my dad and best friend that I now feel as though I wasn't strong enough for *him*. He was so strong for me and my mom. He had to see us cry, plead, and sad much of the time. He was the one dying, yet he was the one trying to cheer *us* up. I feel selfish for that and wish I would have shown him more acceptance and emotional support. Don't get me wrong, we were there for him in every way possible, but in times of fear and weakness, he was the one who was strong for us. Who did he have in his moments of emotional frailty? I know towards the end I did that more for him but wish I would have sooner because I think it hurt him so see us so sad and in denial. Seeing our tears made his heart break and made him worry about us more than he should have. In the end, I let my dad know that he raised a daughter who can hold her head up high and will carry on with his memory and the wonderful lessons and experiences that he raised his family with. I know when he left this world he left it much more at peace, knowing he did his job and that his replacements will continue to make memories like he made with us. Then you will smile, like I can finally do now after three years. If your loved one has to leave this world seeing you a mess, weak, falling apart, and needy, I would think it would be extremely hard on them. They know it sucks, they know you don't want them to leave you, they know you are hurting. But you want to make them proud in life and in death. I think it's easier on them—for sure! Even though my dad is now gone, it's important to continue to make him proud. They would want us to enjoy life *now* and in our days ahead as we would want our loved ones to do if we would have been the one who had died. Go make new memories with them in your heart, and feel the love that is still there and always will be. Smile."

Aaron

"Frayda survived for a period of fifteen months after her diagnosis of metastasized pancreatic cancer even though the original prognosis was three months. We feel that the one-year extension can be attributed to the efficacy of the experimental monoclonal antibody.

"The quality of her life during the first twelve months, while only fair, was good enough to permit some travel during chemo breaks and to obtain fairly frequent visits from family and friends. It was only in the last three months that her condition deteriorated rapidly. She developed pulmonary hypertension, apparently from the Gemzar that was administered along with the monoclonal antibody as part of the phase II study. She had to be removed from the study, and what followed was a series of failed attempts to find another effective treatment.

"When her last CT scan showed extensive spread of the cancer, it was decided by me, in conjunction with her oncologist, to stop the curative treatments, which were completely ineffective, and start palliative care. She was transferred to the hospice wing, where she was kept comfortable but lasted only about three days.

"In discussing the change in care with my children, it was very clear that there was no other choice than the one we took. They were very upset but, at the same time, quite supportive.

"When it was clear that hope was fading, I made arrangements for the funeral without discussing it with Frayda.

"After she passed away, we had an outdoor funeral near the gravesite, which was attended by a large number of friends and relatives.

"For a few days afterwards, I had open house for visitors at my home, and it wasn't until the visits were over that I began to realize the full extent of my loss. After living with someone you've loved for fifty-one years, the house felt very empty, and in a strange way, I still expected her to walk in through the front door.

"Friends and relatives were helpful, but when I went out with them, I had a strong feeling of being lonely, even in a crowd.

"One of the biggest questions that one faces in a situation like this is 'Why fight for additional time when the nature of the disease does not leave room for much (or even any) hope?' The answer to this is that the extra time can be very precious, except, of course, if the patient cannot be kept relatively comfortable. Having a chance to review the good times, even occasionally laugh together and ultimately to say good-bye represents a good way to ease the pain of the transition.

"I get a certain amount of solace from the feeling that we did all we could to keep her as comfortable as possible and provide the best of treatment, even though ultimately it was to no avail. Frayda showed her strength throughout the course of the disease by aiding the oncologist in the choice of the treatment regimen and never complaining or saying anything like 'Why me?'

"In general, cancer treatment and, in some cases, cure, is still in its infancy. Remarkable progress has been made against certain cancers but, unfortunately, not in the case of stage IV pancreatic cancer. Although I don't recommend giving up early in the disease, you have to be prepared for the eventual loss of your loved one.

"If it happens, I would suggest that you dust yourself off and make plans for the rest of your life. Ignore those that say to you that you have not yet grieved enough. The worst thing you can do to yourself is to curl yourself up into the fetal position and remain in your home."

Jennifer

Never Lose Hope

"There are so many things I learned on this journey, and you will too. The first one would be that knowledge is power. Yes, the statistics are scary and daunting, but you or your loved one with pancreatic cancer is different from everyone else. Always keep that in mind. What works for one person does not work for others. Everyone has a different biological makeup.

"Never let go of hope. One day, you will see that it all has finally come together. What you have always wished for has finally come to be. You will look back and laugh at what has passed and you will ask yourself 'How did I get through all of that?

"Hope is always available to us. When we feel defeated, we need only take a deep breath and say 'Yes,' and hope will reappear (Monroe Forester).

"There are people in stage II that die and people in stage IV that go into remission. We do not know why. Remember, a stage is a number, not a death sentence. You are a statistic of one. Always keep that in mind."

Listen and Learn

Find any pancreatic cancer advocacy group you can and read their Web sites. Pancreatic Cancer Action Network has a wonderful program for families and those newly diagnosed. If you are newly diagnosed, you can call and get lots of information and support. The Lustgarten Foundation is another organization that offers much information. There are many others too numerous to mention.

Search *pancreatic cancer* and *pancreatic cancer treatment* on an Internet search engine. Search for clinical trials, treatment, and hospitals or facilities that specialize in pancreatic cancer. If you or your family member is eligible for the Whipple procedure, find someone that has done lots of them. It is best to find a doctor that does only Whipple surgery. It is a complicated and tedious surgery, and you will need to find the best of the best.

Don't be afraid to travel to find the right doctor. There are ways that housing can be found and networks to help you find your way.

I did extensive research and set up a daily Google search e-mail on pancreatic cancer, getting any and all mentions of pancreatic cancer in the media.

Advocate

"The patient must have an active and very supportive advocate. The advocate would be best suited for someone who lives with the patient but could also be a friend. The advocate must accompany the patient to all doctor appointments and procedures. The advocate (with the help of the patient if possible) should keep a daily journal that notes medication taken daily, reactions to medications, food (what you eat each day and how you react to it), bowel status, sleep patterns, pain, and any physical and mental feelings should also be documented in this journal. Extensive record keeping is essential in knowing when certain symptoms change, regress, progress, or disappear.

"Join the Johns Hopkins pancreatic cancer LISTSERV. I found it to be a wonderful resource to help vent, ask questions, and in my case, grieve on their bereavement board. There are others through various areas that can be found on the Internet."

Never Be Afraid to Ask Questions

"Never be afraid to ask a doctor a question. Come to each appointment prepared with a list of questions. That way, you will be able to get them answered or the PA (physician's assistant) can be sure to get them answered for you. Remember that a doctor puts his pants on just like you, one leg at a time. If you don't connect with your doctor or are not comfortable, consider finding another one.

"Do you know of anyone (friends, family, friends of friends, coworkers, etc.) who have any firsthand experience with pancreatic cancer? If so, contact them right away. Networking is essential in finding resources, doctors, medical facilities, and treatments."

Caregivers

"Give yourselves a break. It is a long road—often bumpy, sad, and always life altering. However, you need to take care of yourself so you can be there for the person you are advocating or caring for. There are support groups for caregivers, which I also encourage you to find out about online."

Lastly

"Know that whatever happens, your life will never be the same. You will now likely think of your life as before pancreatic cancer and after pancreatic cancer. Your life and those you love will define and learn a new normal. This is different for everyone. It takes time. Just remember, though, that life is always changing, and we, as mere humans, have little control over much of change in life.

"It is okay to be sad, scared, cry, scream, and vent, whatever you need to do. People who have not experienced it simply don't understand. You can't make them understand. You can only control yourself and your actions. You can find supportive friends even if they have not walked the road that you walk.

"If nothing else, I have learned how to appreciate life with a new and unique philosophy. I appreciate the small things every day as much as I can. Beautiful moments. Silence. Hugs. Love. The importance of family (blood or by choice) and friends that become family.

"Nothing is certain in life but death and taxes. I came to realize that some people survive, and some don't. That is just reality. No one can control this cancer and who will live or die. I also came to understand that when we want someone to live so badly that it can sometimes be harder for them to stay on earth than it is for them to go. You will figure this out if that is the path it takes for you. You will also be grateful for all the moments spent with those you love. You will come to appreciate your loved ones more and relish the good times even while the person may be not feeling so good.

"Be gentle with yourself, and take time for whatever you need to do. Know there are people that care. Wishing you peace, comfort, and strength on your journey, and praying you are one that defies all odds!

"Life is but a stopping place—a pause in what's to be, a resting place along the road to sweet eternity. We all have different journeys, different paths along the way. We all are meant to learn some things but never meant to stay. Our destination is a place, far greater than we know. For some, the journey's quicker. For some, the journey's slow. And when the journey finally ends, we'll claim a great reward and find an everlasting peace."

Sherri

"Here's my advice: be informed, and ask questions. Understand what you are being told, even if that means having it repeated more than once. Have test reports and alternative options explained to you so that you can make an informed decision as to what treatment you decide to undergo. When you go to the doctors initially (or even long term), take a notepad and something to write with and write down what you are being told as you are hearing it. If I did not do this, I would have only remembered about 10 percent of what we were told, and my mother would have remembered less than that. You are hearing a lot of information that you hoped you would never have to hear, and that is like being spoken in a foreign language. It is not expected that you will understand everything you are told the first time, so make notes and educate yourself. Allow yourself time for this to sink in.

"Talk to your doctors, family members, and friends openly and honestly. Your doctors will guide you through what treatment options there may be, but it is ultimately your decision as to what path you take. Remember, if you have been told you have pancreatic cancer, nobody can really tell you anything any worse or that you probably have not thought of already. Be honest with family and friends. They probably do not know what to say to you and would benefit from hearing your openness to discussion. Openness and honesty will also provide you with peace in knowing you have said what you wanted to say without regret and also that others can say those things to you. It gives you both the opportunity to express yourself.

"Don't hide your feelings or keep them bottled up. If you feel like crying, cry, and if you feel like laughing, laugh. At this point, you have very little control over anything, so you might as well have it over yourself. Understand that you are entitled to laugh or cry or fly to the moon if you want to, and nobody has the right to tell you otherwise. This is your life, and if you have not realized this by now, this is the time for you to live it to the fullest as you are able.

"Lastly, try to find humor in everyday things as it is often the little things that provide us with a sliver or glimpse of sanity (something you may not have known for a while). You are hearing so many negative things coming at you from every direction that the ability to find humor in something, anything, is a rare treat in itself."

Deborah

"One of the most significant things I've learned from this experience is how important it is to be cognizant of your family history. Before my mother's diagnosis, I had no idea that three out of four of my grandparents had died of cancer, with two of them dying of pancreatic cancer. If we had been aware of this, we may have been more assertive about screening or more concerned about the seemingly minor symptoms my mother exhibited months before diagnosis. I would strongly suggest that everyone draw out a family tree to trace genetic diseases and pay particular attention if history suggests you need to."

Michele

"My dad was a great storyteller, and I wish I would have recorded him telling stories and jokes. Although they wouldn't have been the same because he was sick, I treasure the four voice mails I kept and even had them burned to a CD so I could make sure to hear his voice forever.

"I would talk a lot about favorite times and experiences you had together. I asked my dad a lot of questions when he was feeling up to it when he was battling, and I wish I would have asked more. Also, write everything down. I wrote some journal entries at the time but wished I would have more.

"Watch old home movies of when you were little. I never did this and still haven't been able to turn them on, but I know that those were the best times for my dad, and it would have been nice to have him see those again.

"Hold hands, and touch them as much as you can. I remember looking at my dad's hand and thinking that I loved them so much. Over the years, I watched those hands raise me, and all I wanted to do was hold on to them.

"Don't spend too much time on the Internet looking for cures and new treatments, etc. Find the best doctors and listen to them. Always ask as many questions as you think of and write down the answers. Get second opinions if you need to, but just realize that the time you spend with the person is the most important thing.

"Don't get mad if your patient wants to drink wine or eat sugar or whatever. This may be a coping mechanism that, at this time, may actually help relieve stress rather than make the problem worse. Just be supportive, and allow them to indulge. Obviously not overboard, but don't fight them on it. My dad ordered a beer at a restaurant, and I got so upset. Another time, he ordered his favorite cold sake at a sushi restaurant. I said, 'Dad, let's just get you through this and then you can drink.' He said, 'What if it never goes away? What if I have this forever?' That sunk in. He just wants to be him. Let them be them. No need to try to change them now. It will just cause fights. Just *love* them!

"When you are in the hospital, fight and advocate for your patient. Make good friends with the good nurses. Understand what they are doing to make sure the next nurse does it right. If you feel like something is not right, ask for the nursing manager. Do not be afraid to advocate for your patient. If you don't, no one will. There were two times in my dad's sickness where they forgot to put him on an IV drip when his fluids were draining. He got so dehydrated that rapid response had to be called. The second time was when he was internally bleeding, and it caused the chaos that inevitably ended his life. Be vigilant, have all the info with you at all times, ask questions.

Say I love you more than you need to through actions and words. I told my dad I loved him so many times that he sometimes would respond with 'I know,' not 'I love you too!' I jokingly got mad, but I needed him to know!

"Spend as much time with them as possible. Although not everyone has the luxury of being able to quit their job and take over full-time care of their patient, I did. I do not regret one moment. I met a nurse when we were in the first hospital stay, and she gave me a binder to start keeping things together. She told me that she took care of both her parents when they were sick, and she didn't regret one moment. I would have done it anyway, but it hit me then how serious this illness was and that every moment would be important. It was, and I am so thankful that I could! I would make him smoothies every day and watch movies on the couch and take him to all his appointments. I miss all those moments even though some were the hardest of my life."

Lisa

"As I sit outside on my patio on this beautiful spring day, I am filled with so many emotions. I think about the life I thought that Fred and I would have together and the dreams we had planned. I think about the thirty-three years together. He was my partner, my best friend, my lover, and a wonderful, devoted father to our children. I think about how unbelievable it is that I am in this position, trying to rebuild my life and making decisions alone.

"When I think back about the fifteen months that Fred had since his diagnosis, I remember the constant pain that he was in. I also think about how unfair it was that this diagnosis came at a point in his life that he was moving along with his career and that we had so much to look forward to in our lives.

"Now, in my second year without Fred, I try not to think about what-ifs. I realize that as the patient, Fred did what he felt he could live with. As the caregiver, my role was to support him, give him unconditional love, and be there for him. Who knows what I would have done if I were the patient? I tried to find support through family, friends, and groups. I did constant research to make me feel that I was doing everything I could possibly do. However, in the end, you come to realize that it is not about your life—it is about the patient's. As difficult as it is, you have to stand beside your loved one. You can give all the advice you want, but in the end, the patient has the ultimate right to decide about the life he/she wants to lead. Probably the most precious form of love is supporting that person in his decisions, though you may disagree.

"There isn't a day that goes by that I don't think of Fred. I keep pictures up, and I talk about him. I am so grateful for all the years we spent together. I am so sad for not having more. I continue to add to Fred's online memorial Web site. It brings me comfort. I am constantly reminded that life goes on. I have celebrated two birthdays and two wedding anniversaries, two Mother's Days, and two Father's Days without Fred. I celebrated my son's twenty-first birthday and made it a really special day. I have my daughter's wedding coming up in the fall. I am selling my motor home. I have paid off my house and any loans we had.

"I don't know what lies ahead, but I will reiterate Fred and my philosophy of life. We didn't need pancreatic cancer to remind us about what was important. Live life to the fullest, and know what is important. This is not a dress rehearsal."

MEDICAL GLOSSARY*

Ascites — excess fluid in the space between the membranes lining the abdomen and abdominal organs.

Adenocarcinoma –a cancer of epithelia originating in glandular tissue. Epithelial tissue includes, but is not limited to, skin, glands and a variety of other tissue that lines the cavities and organs of the body.

Biliary drain — A biliary drain is a tube to drain bile from your liver. It is put in by a doctor called an Interventional Radiologist. The tube or catheter is placed through your skin and into your liver. You may also hear your drain called a biliary stent or biliary catheter.

CA 19-9 — This is a blood test that is useful in assisting with initial diagnosis. The higher the CA19-9 level, the larger the tumor and the less chance that the tumor can easily be cut back. In evaluating treatments, a decreasing or stable CA19-9 level generally indicates an improved chance of survival, while an increasing level indicates the progression of disease.

CT scan — also called computerized tomography or just CT—combines a series of X-ray views taken from many different angles to produce cross-sectional images of the bones and soft tissues inside your body

CyberKnife — The CyberKnife® Robotic Radiosurgery System is a non-invasive alternative to surgery for the treatment of both cancerous and non-cancerous tumors anywhere in the body, including the prostate, lung, brain, spine, liver, pancreas and kidney. The treatment—which delivers high doses of radiation to tumors with extreme accuracy—offers new hope to patients who have inoperable or surgically complex tumors, or who may be looking for a non-surgical option.

Endoscopic Retrograde Cholangio Pancreatography (ERCP) — Endoscopic retrograde cholangiopancreatography (ERCP) is a technique that combines the use of endoscopy and fluoroscopy to diagnose and treat certain problems of the biliary or pancreatic ductal systems. Through the endoscope, the physician can see the inside of the stomach and duodenum, and inject dyes into the ducts in the biliary tree and pancreas so they can be seen on x-rays. (Wikipedia)

Endoscopic ultrasound (EUS) — echo-endoscopy is a medical procedure in endoscopy (insertion of a probe into a hollow organ) is combined with ultrasound to obtain images of the internal organs in the chest and abdomen. It can be used to visualize the wall of these organs, or to look at adjacent structures. Combined with Doppler imaging, nearby blood vessels can also be evaluated.

FDA Approval — The **Food and Drug Administration** (**FDA** or **USFDA**) is an agency of the United States Department of Health and Human Services, one of the United States federal executive departments, responsible for protecting and promoting public health through the regulation and supervision of food safety, tobacco products, dietary supplements, prescription and over-the-counter pharmaceutical drugs (medications), vaccines, biopharmaceuticals, blood transfusions, medical devices, electromagnetic radiation emitting devices (ERED), veterinary products, and cosmetics.

Gastroenterology — Gastroenterology (MeSH heading)[1] is the branch of medicine whereby the digestive system and its disorders are studied. Etymologically, the name is a combination of three Ancient Greek words gaster (gen.: gastros) (stomach), enteron (intestine), and logos (reason).

Gemzar — Gemcitabine (pronunciation: jem-SITE-a-been) is a nucleoside analog used as chemotherapy. It is marketed as Gemzar by Eli Lilly and Company.

Gemzar Taxotere, and Xeloda. (GTX) — for pancreatic cancer — GTX is a promising regimen for pancreatic cancer but still not adequately supported.

Current treatment options for pancreatic cancer include surgery, radiation, and chemotherapy. Although single-agent chemotherapy with gemcitabine has been a standard treatment for pancreatic cancer, research has increasingly focused on the development of combination regimens. Recently, the U.S. Food and Drug Administration approved the combination of the targeted therapy Tarceva ® (erlotinib) and gemcitabine for the treatment of locally advanced, inoperable, or metastatic pancreatic cancer

GERD — Gastroesophageal reflux disease (GERD) is a condition in which the stomach contents (food or liquid) leak backwards from the stomach into the esophagus (the tube from the mouth to the stomach). This action can irritate the esophagus, causing heartburn and other symptoms.

Gastrojejunostomy — A Gastrojejunostomy is performed by inserting a tube directly into the stomach to the small intestine of a person who cannot take food or medicine by mouth. Gastrojejunostomy can be done surgically, or through an Interventional Radiology (IR) technique called "Percutaneous Gastrojejunostomy," which requires only a tiny incision in the skin. Percutaneous Gastrojejunostomy can be performed safely in adults and children. Generally this involves a short hospital stay.

Hemochromatosis — Hemochromatosis is a disorder that interferes with the body's ability to break down iron, and results in too much iron being absorbed from the gastrointestinal tract.

Hospice — a type of care and a philosophy of care which focuses on the palliation of a terminally ill patient's symptoms. These symptoms can be physical, emotional, spiritual or social in nature.

IV — Intravenous therapy or IV therapy is the giving of liquid substances directly into a vein. The word intravenous simply means "within a vein". Therapies administered intravenously are often called specialty pharmaceuticals. It is commonly referred to as a drip because many systems of administration employ a drip chamber, which prevents air entering the blood stream (air embolism) and allows an estimate of flow rate.

Monoclonal Antibody — Monoclonal antibodies (mAb or moAb) are monospecific antibodies that are the same because they are made by one type of immune cell that are all clones of a unique parent cell. Given almost any substance, it is possible to create monoclonal antibodies that specifically bind to that substance; they can then serve to detect or purify that substance. This has become an important tool in biochemistry, molecular biology and medicine. When used as medications, the non-proprietary drug name ends in—mab (see "Nomenclature of monoclonal antibodies").

MRI — Magnetic resonance imaging (MRI), or nuclear magnetic resonance imaging (NMRI), is primarily a medical imaging technique used in radiology to visualize detailed internal structure and limited function of the body. MRI provides much greater contrast between the different soft tissues of the body than computed tomography (CT) does, making it especially useful in neurological (brain), musculoskeletal, cardiovascular, and oncological (cancer) imaging. Unlike CT, MRI uses no ionizing radiation. Rather, it uses a powerful magnetic field to align the nuclear magnetization of (usually) hydrogen atoms in water in the body. Radio frequency (RF) fields are used to systematically alter the alignment of this magnetization. This causes the hydrogen nuclei to produce a rotating magnetic field detectable by the scanner. This signal can be manipulated by additional magnetic fields to build up enough information to construct an image of the body.

Neuroendocrine carcinoma — Neuroendocrine tumors, or more properly gastro-entero-pancreatic or gastroenteropancreatic neuroendocrine tumors (GEP-NETs), are cancers of the interface between the endocrine (hormonal) system and the nervous system.

Paracentesis — a medical procedure involving needle drainage of fluid from a body cavity, most commonly the peritoneal cavity in the abdomen.

Pancreatitis — inflammation of the pancreas that can occur in two very different forms. Acute pancreatitis is sudden while chronic pancreatitis "is characterized by recurring or persistent abdominal pain with or without steatorrhea or diabetes mellitus.

PET scan — Positron emission tomography (PET) is a nuclear medicine imaging technique which produces a three-dimensional image or picture of functional processes in the body. The system detects pairs of gamma rays emitted indirectly by a positron-emitting radionuclide (tracer), which is introduced into the body on a biologically active molecule. Images of tracer concentration in 3-dimensional or 4-dimensional space (the 4th dimension being time) within the body are then reconstructed by computer analysis. In modern scanners, this reconstruction is often accomplished with the aid of a CT X-ray scan performed on the patient during the same session, in the same machine.

Phase I clinical trial — Initial studies to determine the metabolism and pharmacologic actions of drugs in humans, the side effects associated with increasing doses, and to gain early evidence of effectiveness; may include healthy participants and/or patients.

Pulmonary Hypertension — Pulmonary hypertension is abnormally high blood pressure in the arteries of the lungs. It makes the right side of the heart need to work harder than normal.

Stage III-IV — clinical trials of new drugs are usually administered by a contract research organization (CRO) hired by the sponsoring company . . .

Stage IV Pancreatic Cancer — Pancreatic cancer is considered Stage IV if it has spread to distant locations in the body, such as the liver, lungs, or adjacent organs including the stomach, spleen, and/or the bowel. Sometimes it can only be determined that a pancreatic cancer is in Stage IV once surgery is completed.

Tarceva — Tarceva® (erlotinib) is a small molecule human epidermal growth factor type 1/epidermal growth factor receptor (HER1/EGFR) inhibitor which demonstrated, in a Phase III clinical trial, an increased survival in advanced non-small cell lung cancer (NSCLC) patients. In a Phase III trial, Tarceva has also shown an improvement in overall survival when added to gemcitabine chemotherapy as initial treatment for advanced pancreatic cancer.

Taxotere — the trade name for the generic chemotherapy drug Docetaxel. **Taxotere** is an anti-cancer (antineoplastic or cytotoxic) chemotherapy drug.

Whipple — In the Whipple operation the head of the pancreas, a portion of the bile duct, the gallbladder and the duodenum is removed. Occasionally a portion of the stomach may also be removed. After removal of these structures the remaining pancreas, bile duct and the intestine is sutured back into the intestine to direct the gastrointestinal secretions back into the gut.

Xeloda — the trade name for the generic chemotherapy drug Capecitabine. **Xeloda** is an anti-cancer chemotherapy drug is classified as an antimetabolite.

*References from Mayo Clinic, Wikispaces, WebMD

AUTHOR BIOGRAPHY

Born in Brooklyn, New York, and raised as an only child on Long Island, Lisa M. Strahs-Lorenc (maiden name Lisa Melanie Strahs) always had a positive outlook in life, even though her biological mother died when she was two years old (her father remarried when she was four and a half) and her father died when she was thirty. Because of the losses in her life, she always believed in living life to the fullest and making the most of each day.

She had always dreamed of being an early-childhood educator. She arrived at Syracuse University in the fall of 1972 and was introduced through her roommate to Fred Lorenc (from Cleveland, Ohio), who was studying architecture. Timing is everything, and on September 3, 1975, both were ready for a real relationship. Settling in Waltham, Massachusetts, Lisa began her short-lived early-childhood career at a nursery school, and Fred started in an architecture firm. Her next job, after the school went bankrupt, was at an international adoptions agency as the office manager. In the meantime, Fred continued to grow with the architecture firm and excel in his profession.

They were married on May 28, 1978, and had their names legally hyphenated, after Fred stated that he couldn't understand why a woman should have to lose her name. Only one year later, they decided to make the move to Long Island to be near Lisa's parents. Both found jobs in New York City. Fred found another architecture firm, and Lisa found a children's fund-raising organization for blood diseases. After receiving her master's degree with a specialization in career counseling (she felt that she did not receive the kind of guidance that she would have liked when she was in high school), she worked at a psychotherapy office, which also had a teaching institute. In the meantime, Fred too left the city and worked for several architecture firms. By now, they had been together for almost nine years.

One never knows what life will bring, and that was so apparent the night that Lisa and Fred received a call from Toronto, Canada, from a hospital nurse that stated that Lisa's father had had a heart attack and died. Lisa flew back with her mom and her father's body while Fred drove their car back to Long Island. Lisa went home to write a eulogy that she would deliver the next day. She was thirty years old, and her father was fifty-seven when he died on May 23, 1984. Knowing that there could never be a replacement for her father, she also felt that life needed to go on and a child would bring new life to the family.

On November 11, 1985, Sari Allison Strahs-Lorenc was born. She was named after Lisa's father, Sigmund, and her biological mother, Adele. Lisa also attended a bereavement group and contributed to a book entitled *Recovering from the Loss of a Parent*. After her birth, Lisa began a typing service. In addition, for a couple of hours a week, she worked as an adjunct career counselor at a local community college. In the meantime, Fred continued to grow in his profession, though it became clear that architecture was starting to become a dying trade. It was his income that allowed Lisa to start her business and follow her dreams.

Lisa had always wanted more than one child, so on May 25, 1989, Ryan Francis Strahs-Lorenc was born. He was named after Fred's father, Ferdinand, and Lisa's grandfather Reuben. Lisa started a home-based day care center and a career-counseling service while Ryan was young. As the children got older, Lisa worked as the marketing manager for a couple of software companies. Then Lisa wrote her job description for a nonprofit organization that builds partnerships between the academic and business communities to prepare students with careers for the twenty-first century. She has worked for this organization for almost eleven years.

In the meantime, Lisa encouraged Fred to attend law school at night so that he could pursue his own dreams. He became an accomplished lawyer, though starting from the bottom and working his way up again wasn't easy. After a couple of years, fate struck again, and after a thyroidectomy, he began to complain of serious stomach pains. Only four months later, Fred was diagnosed with pancreatic cancer. He lived for fifteen months, always in pain but fighting the disease with radiation and chemotherapy. During the entire illness, Lisa was his advocate and fellow warrior, seeking opinions from leading cancer treatment centers and doing research on the Internet. After his death (she delivered a very personal eulogy at his memorial service), she created an online memorial Web site in his memory.

She continues to write in the journal and encourage visitors to leave virtual flowers and candles and to sign the guestbook. During the illness and after, she continued to seek out support and bereavement groups. It is through this experience that inspired her to write this book. After significant research, it was apparent that there were no support books for families and patients with this deadly disease. Lisa wanted to gather stories, feelings, and advice from patients, spouses, siblings, and adult children that would help others and raise money for two organizations that offer support and research toward an eventual cure. The proceeds of this book are being donated to the Lustgarten Foundation and the Pancreatic Cancer Action Network.

Fred's memory is always with Lisa through her daily activities and the life she continues to lead. As difficult as it is, life does go on. It will certainly never be the same, but Lisa still believes today that you need to live life to the fullest and enjoy each day no matter what the future holds.

www.ingramcontent.com/pod-product-compliance
Lightning Source LLC
Chambersburg PA
CBHW020421290526
45785CB00002B/674